普通高等教育"十三五"规划教材
高职高专实验（训）系列

PHP 实训教程

主　审　张小红
主　编　马　杰
副主编　罗东芳
　　　　炊　昆

立信会计出版社
LIXIN ACCOUNTING PUBLISHING HOUSE

图书在版编目(CIP)数据

PHP实训教程 / 马杰主编. —上海：立信会计出版社，2017.1
普通高等教育"十三五"规划教材
ISBN 978－7－5429－5314－8

Ⅰ.①P… Ⅱ.①马… Ⅲ.①PHP语言—程序设计—高等学校—教材 Ⅳ.①TP312

中国版本图书馆CIP数据核字(2016)第325968号

责任编辑　赵新民
封面设计　南房间

PHP 实训教程
PHP Shixun Jiaocheng

出版发行	立信会计出版社			
地　　址	上海市中山西路2230号	邮政编码	200235	
电　　话	(021)64411389	传　　真	(021)64411325	
网　　址	www.lixinaph.com	电子邮箱	lxaph@sh163.net	
网上书店	www.shlx.net	电　　话	(021)64411071	
经　　销	各地新华书店			
印　　刷	江苏凤凰数码印务有限公司			
开　　本	787毫米×1 092毫米	1/16		
印　　张	16			
字　　数	420千字			
版　　次	2017年1月第1版			
印　　次	2017年1月第1次			
书　　号	ISBN 978－7－5429－5314－8/TP			
定　　价	36.00元			

如有印订差错,请与本社联系调换

普通高等教育"十三五"规划教材

高职高专实验(训)系列

编委会主任　赵水根

编委会副主任　王振华　张学功

编 委 会 委 员　（以姓氏笔画为序）
　　　　　　　　马荣贵　孔祥慧　宁艳岩　刘爱萍　刘　喆
　　　　　　　　张效梅　李煜辉　陈爱国　倪天林　琚军红
　　　　　　　　董云展　韩宗保

　　　　　　　　行业企业委员　（以姓氏笔画为序）
　　　　　　　　王寿轩　牛宗芬　史　强　石维堂　张连升
　　　　　　　　张延民　赵永战　赵树亭　臧喜昌

总 序 PREFACE

实验(训)教学是高等职业教育教学的重要环节,是培养适应现代经济社会发展的高素质技能人才的重要保障。规范实验(训)教学内容,建立标准化的实验(训)教学流程是完善实践教学体系,推进人才培养规范化,加快发展现代职业教育的重要举措。为此,我们编纂了本套实验(训)系列教材。

本系列教材在编纂过程中,紧密结合行业企业发展实际,坚持应用导向,坚持实践教学与理论教学相衔接,实践内容与职业标准相衔接,实践技能与职业技能鉴定相衔接,把职业岗位所需要的知识、技能和职业素养融入实践教学,构建对接紧密、特色鲜明的实践教学课程体系。

本系列教材在栏目编排上,采用模块化的结构,系统讲解实践教学的各个环节。同时,本系列教材紧贴实践教学内容,采用项目教学、案例教学、工作过程导向教学等教学模式。

为确保教材质量,本系列教材由具有企业一线工作经历和丰富实践教学经验的"双师型"教师编写。在写作方式上,本系列教材力求语言简练、形式活泼、深入浅出。本系列教材以课程为单元,配有丰富的实验(训)案例,是高校教师教授实践类课程的重要参考。

普通高等教育"十三五"规划教材编委会

前言 FOREWORD

PHP 具有简单易学、开源、高效、较高安全性、运行环境成本低等优点，在近几年已经成为主流开发语言，并长期高居 TIOBE 语言排行榜前十位。Facebook、雅虎、百度、维基百科、腾讯、淘宝、新浪等网站前端均采用 PHP 作为开发语言。

随着 PHP 的流行，可以看到，各个企业开始大量招收 PHP 相关的 Web 开发人才，各个培训机构也将 PHP 培训纳入重点培训项目中，国内高校也开始陆续开设 PHP 开发课程。但是，传统的课堂教学，注重理论，无法适应现在的市场需求，很难培养出动手能力很强的优秀毕业生，从而造成了 PHP 开发人才的大量空缺。

为了解决课堂实训教学不足的问题，也为了方便读者快速学习 PHP 程序开发，本书将重点放在实训上。通过由浅入深，由简到难的实训，再搭配完整的项目开发，让读者能够轻松入门，搭建出自己满意的 PHP 网站。

本书一共分为三部分：基础篇、拓展提高篇、项目实战篇。

1. 基础篇主要进行 PHP 相关基础知识的实训，包括 Html、DIV＋CSS、JavaScript、Apache＋PHP 服务器配置、PHP 语言基础、PHP 面向对象、PHP 内部函数、MySQL 服务配置等相关知识。

2. 拓展提高篇主要进行目前市场上 PHP 的一些主流应用技术的实训，包括 PHP 数据库基本操作、PHP 数据库查询分页技术、用户登录控制、用户权限管理、FCKeditor、数据导入 office、jQuery 应用等实用技术。

3. 项目实战篇通过 3 个不同项目的开发，进行实战综合实训。包括教务管理系统开发、库存管理系统开发以及使用帝国 CMS 进行的公司网站开发。

本书的内容包括 PHP 网站制作技术的各个方面，通过大量实训，加强了读者对 PHP 理论的理解，提高了实际动手能力，有助于毕业生快速融入工作岗位。本书适合作为高校理论教材的配套实训教材使用，也可以单独用于有一定 PHP 基础的读者进行实训提高。

本书由马杰任主编，罗东芳、炊昆任副主编。其中马杰负责全书内容与结构的规划,并编写项目一至项目三、项目十二至项目十四;罗东芳负责编写项目八至项目十一,炊昆负责编写项目四至项目七,同时感谢张小红老师的大力支持和修改建议,感谢德亿电子科技有限公司郑俊岭女士对本书项目实战部分的指导。

  由于作者水平有限,加之网站开发技术发展迅速,本书的内容难免会有纰漏和不足之处,恳请各位专家、同仁和读者批评指正。

编　　者

目录 CONTENTS

项目一　Dreamweaver 安装使用 ･･･ 1

项目二　JavaScript 应用 ･･ 5
　　任务一　编写网页进度条 ･･･ 5
　　任务二　编写背景自动变色网页 ･･･････････････････････････････････････ 7
　　任务三　编写选项卡效果网页 ･･･ 8
　　任务四　日期计算 ･･･ 10
　　任务五　编写自动隐藏显示网页 ･･･････････････････････････････････････ 12
　　任务六　文字旋转效果 ･･･ 13

项目三　构建动态 Web 开发环境 ･･ 15
　　任务一　安装 Apache,配置网站服务器 ････････････････････････････････ 15
　　任务二　PHP 的安装与 Apache 结合使用 ･･････････････････････････････ 23

项目四　PHP 内部函数应用 ･･･ 26
　　任务一　PHP 时间日期函数 ･･･ 26
　　任务二　PHP 文件函数 ･･･ 26
　　任务三　PHP 正则表达式 ･･･ 27
　　任务四　FTP 函数 ･･･ 28

项目五　MySQL 数据库的安装配置 ･･ 30
　　任务一　安装配置 MySQL 数据 ･･･････････････････････････････････････ 30
　　任务二　构建 Dreamweaver 下的动态站点 ････････････････････････････ 38

项目六　PHP 对 MySQL 数据的基本操作 ･･･････････････････････････････････ 40
　　任务一　信息添加 ･･･ 40
　　任务二　信息修改 ･･･ 43
　　任务三　信息删除 ･･･ 45

项目七　PHP 对 MySQL 数据的查询 ················ 47
　任务一　数据基本查询 ················ 47
　任务二　数据分页显示 ················ 48
　任务三　创建分页类 ················ 51
　任务四　多条件数据查询 ················ 56

项目八　用户登录控制 ················ 60
　任务一　PHP session 的配置 ················ 60
　任务二　基本登录控制 ················ 62
　任务三　多级权限登录控制 ················ 64
　任务四　带验证码的登录控制 ················ 67

项目九　FCKeditor 应用 ················ 70
　任务一　FCKeditor 的下载和配置 ················ 70
　任务二　新闻编辑 ················ 73
　任务三　新闻浏览页面 ················ 76

项目十　数据导入 Office ················ 79
　任务一　数据导入 Excel ················ 79
　任务二　数据导入 Word ················ 88

项目十一　jQuery 应用 ················ 90
　任务一　jQuery 引用 ················ 90
　任务二　jQuery 隐藏显示 ················ 91
　任务三　jQuery 淡入淡出 ················ 92
　任务四　jQuery 滑动效果 ················ 94
　任务五　jQuery 动画 ················ 95
　任务六　jQuery 半透明遮罩效果 ················ 96
　任务七　jQuery 层的拖动 ················ 98
　任务八　jQuery 下拉导航菜单 ················ 99

项目十二　教务管理系统 ················ 110
　任务一　功能结构图 ················ 110
　任务二　数据库设计 ················ 111
　任务三　登录页面 ················ 113
　任务四　学生主页 ················ 117
　任务五　学生信息查询页面 ················ 119
　任务六　教师功能页面 ················ 141
　任务七　管理员功能页面 ················ 147

项目十三 库存管理系统 ... 156
- 任务一 功能模块设计 ... 156
- 任务二 数据库设计 ... 157
- 任务三 登录功能 ... 160
- 任务四 店外商品查询功能 ... 163
- 任务五 店内商品查询及销售 ... 166
- 任务六 库存警报 ... 170
- 任务七 修改商品价格 ... 173
- 任务八 销售情况饼状图 ... 176
- 任务九 新闻管理 ... 177
- 任务十 入库管理 ... 185

项目十四 帝国 CMS ... 195
- 任务一 帝国 CMS 的安装和配置 ... 195
- 任务二 模板管理 ... 199
- 任务三 栏目管理 ... 236
- 任务四 信息管理 ... 240

参考文献 ... 242

项目一 Dreamweaver 安装使用

一、实训目的

- 掌握 Dreamweaver 的安装方法；
- 掌握 Dreamweaver 创建网页的方法。

二、实训要求

- 完成 Dreamweaver 的安装；
- 使用 Dreamweaver 创建设计网页。

三、实训设计

HTML 是创建 Web 页面的语言，如今现有的每个 Web 浏览器都能理解这种语言，而 Dreamweaver 则是页面设计的辅助工具，在业内使用率极高。为了锻炼学生的软件安装能力和页面设计能力，要求学生自行从网络上下载 Dreamweaver 安装包，并进行安装，最后利用 Dreamweaver 设计页面。

四、实训内容

（1）利用搜索引擎，搜索 Dreamweaver，然后下载。

（2）下载后解压，然后点击安装，见图 1.1。

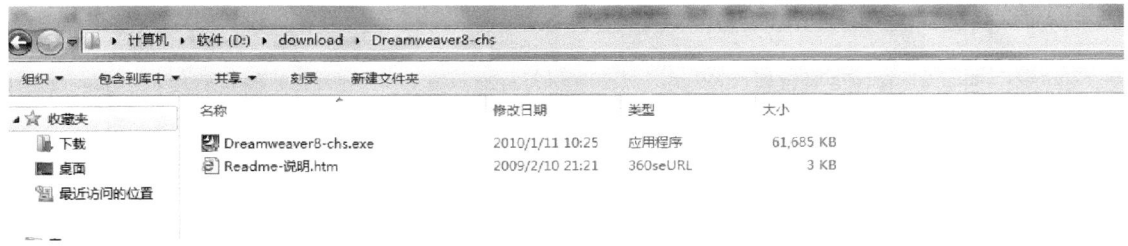

图 1.1 解压缩

（3）按照安装屏幕上的指导执行操作，见图 1.2 至图 1.9。

（4）第一次打开 Dreamweaver 8，在出现工作区的提示项中选择（选设计器），然后选"我有一个序列号/继续/输入序列号"。把注册数列号粘贴在方格里。如果在电脑里已装了网页三剑客的其他两个软件（fireworks flashf）中任意一款软件，此项就没有了，在桌面上直接打开软件就行了。

序列号：
WPD800-51333-38632-90078
WPD800-55537-19732-30159
WPD800-59731-73932-34113
WPD800-55535-30132-25437

图 1.2　欢迎使用

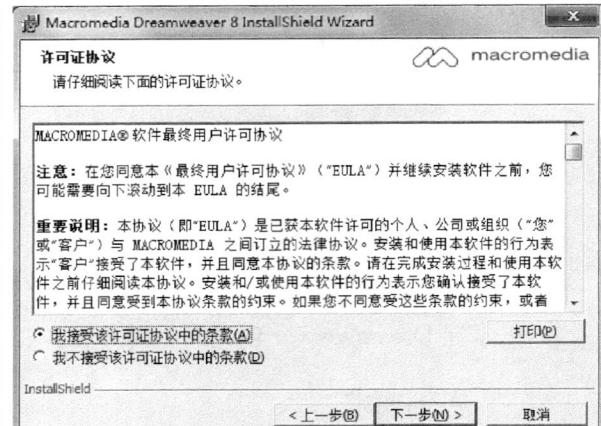

图 1.3　许可协议

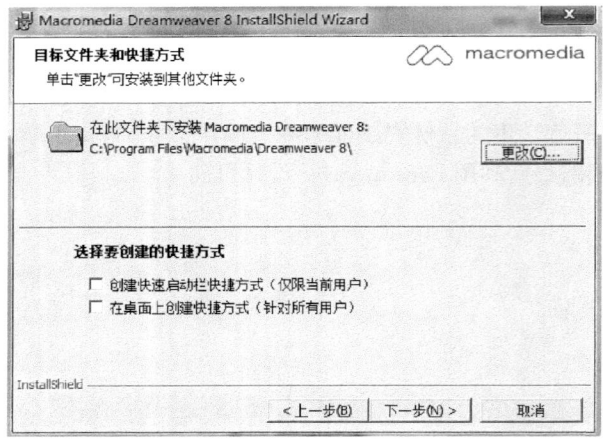

图 1.4　更改安装位置

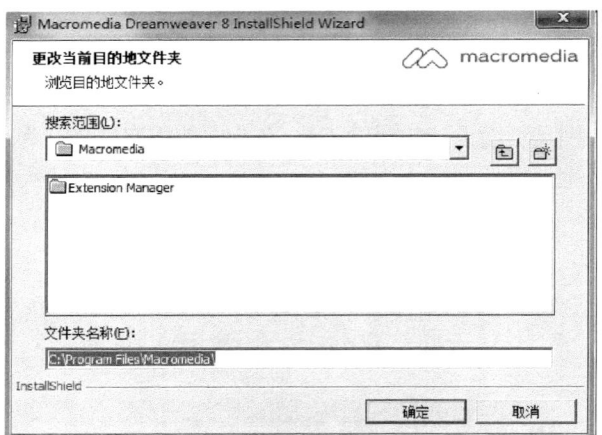

图 1.5　确定更改

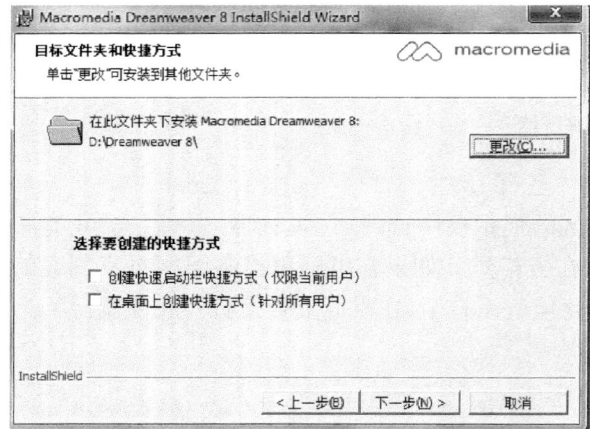

图 1.6　继续安装

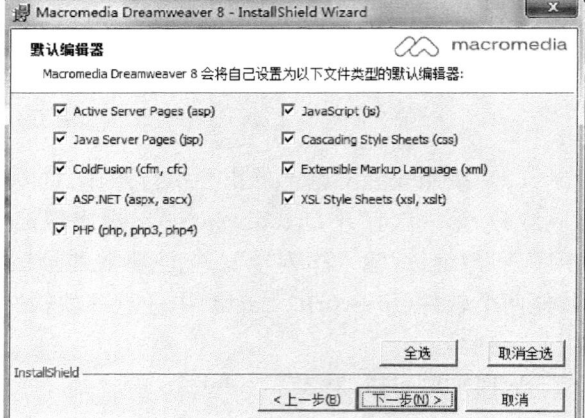

图 1.7　默认编辑器

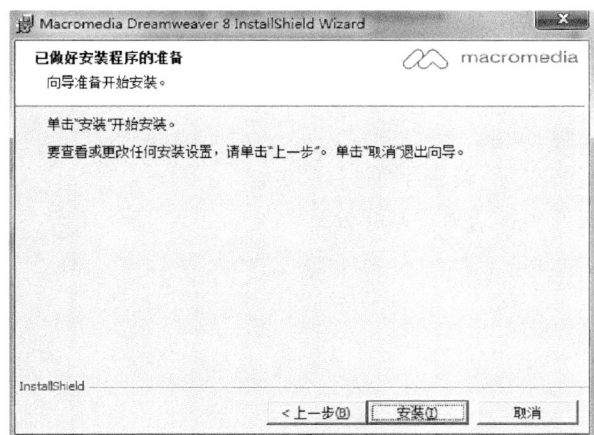
图 1.8 安装

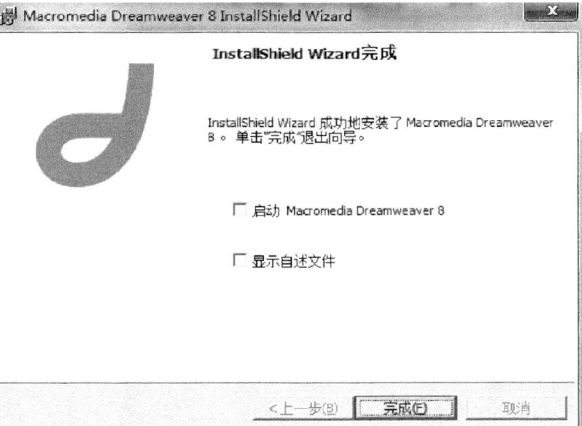
图 1.9 完成安装

(5) 再打开即图 1.10 的界面。

图 1.10 界面图

(6) 点击创建新项目下的 HTML,即可进入工作界面,开始设计网页,内容不限,见图 1.11。

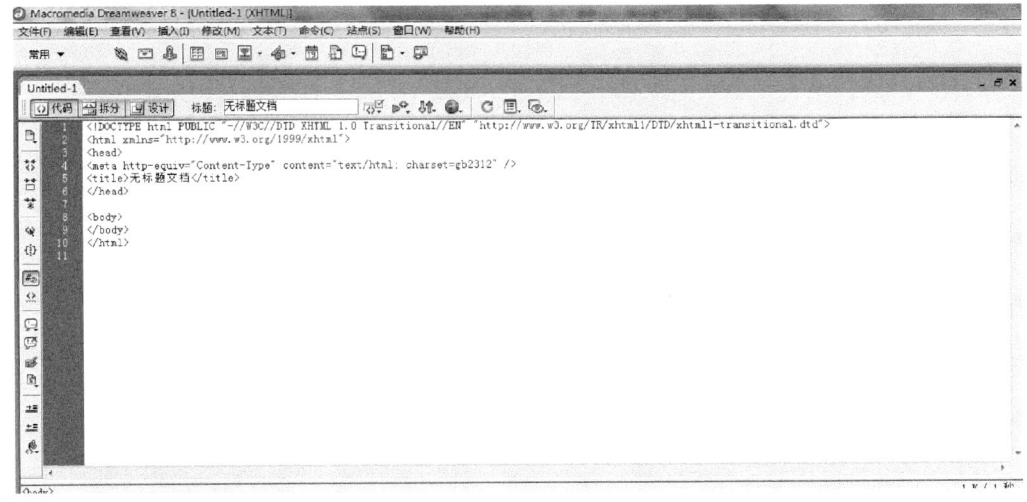

图 1.11 页面设计

五、考核标准

(1) 主题鲜明,具有创意。(10分)

(2) 版面布局合理清晰,整体效果美观,观赏性强。(10分)

(3) 网页中没有明显的错误(如超链接、图片无法显示、错别字等)。(30分)

(4) 项目文档内容详细完整,结构清晰,提交文件名称正确。(20分)

(5) 其他技术。(30分)

项目二　JavaScript 应用

一、实训目的

- 掌握 JavaScript 的基本语法；
- 掌握 JavaScript 动态页面技术。

二、实训要求

- 使用 JavaScript 在网页中添加动态效果；
- 使用 JavaScript 实现对页面元素的动态控制；
- 使用 JavaScript 动态选项卡效果。

三、实训设计

JavaScript 是一种直译式脚本语言，是一种动态类型、弱类型、基于原型的语言，内置支持类型，是网页客户端必不可少的动态脚本语言。为了锻炼学生的页面动态效果制作能力，本实训设计了多个不同的题目，包括页面进度条、改变页面背景颜色、选项卡效果、页面导航等内容。通过不同内容的实训，可以有效提高学生的动手能力，以应对不同的功能需求。

四、实训内容

任务一　编写网页进度条

编写一个网页文件。要求：使用 JavaScript 实现网页上常见的 Loading 效果，如 Windows XP 启动时候的进度条效果，左右来回跑动的彩带，并显示文字提示"程序正在加载中……"。

```html
<html>
<head>
<title>程序加载页面</title>
<meta http-equiv="Content-Type" content="text/html; charset=gb2312">
</head>
<body style="background:black">
<div id="div1" style="position:absolute;width:322;height:14;border:1 #707888 solid;overflow:hidden">
    <div style="position:absolute;top:-1;left:0" id="pimg">
    </div>
</div>
<div id="div2" style="position:absolute;top:30;left:120;font-size:9pt;color:#f4f4f4">
正在加载中……
</div>
<script language="JavaScript">
s=new
```

```
Array("#050626","#0a0b44","#0f1165","#1a1d95","#1c1fa7","#1c20c8","#060cff");
//s=new
Array("#333333","#555555","#777777","#999999","#AAAAAA","#CCCCCC","#EEEEEE");
div1.style.posTop=Math.floor((document.body.clientHeight-14)/2);
div1.style.posLeft=Math.floor((document.body.clientWidth-322)/2);
div2.style.posTop=parseInt(div1.style.posTop)+20;
div2.style.posLeft=parseInt(div1.style.posLeft)+120;
function Larrange(){
    pimg.innerHTML="";
    for(i=0;i<9;i++){
        pimg.innerHTML+="<input style=\"width:15;height:10;border:0;background:"+s[i]+";margin:1\">";
    }
}
function Rarrange(){
    pimg.innerHTML="";
    for(i=9;i>-1;i--){
        pimg.innerHTML+="<input style=\"width:15;height:10;border:0;background:"+s[i]+";margin:1\">";
    }
}
var is=0;size=0;
function move(){
    if(pimg.style.pixelLeft<350&&is==0){
        if(size==0){Larrange();size=1;}
        pimg.style.pixelLeft+=3;
        setTimeout("move()",1);
        return;
    }
    is=1;
    if(pimg.style.pixelLeft>-200&&is==1){
        if(size==1){Rarrange();size=0;}
        pimg.style.pixelLeft-=3;
        setTimeout("move()",1);
        return;
    }
    is=0;
    move();
}
function flashs(){
    if(div2.style.color=="#ffffff"){
        div2.style.color="#707888";
        setTimeout('flashs()',500);
    }
    else{
        div2.style.color="#ffffff";
        setTimeout('flashs()',500);
    }
}
Larrange();
flcshs();
```

```
move();
</script>
</body>
</html>
```

代码效果见图2.1。

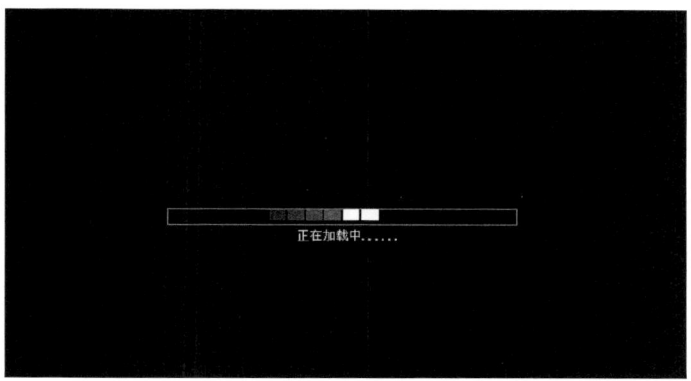

图2.1　进度条

任务二　编写背景自动变色网页

编写一个网页文件，JavaScript实现网页背景自动变色，自己变换颜色，设定时间和颜色值即可，在你设定的颜色值、一定时间内自动切换网页背景颜色。

```
<!DOCTYPE HTML PUBLIC "-//W3C//DTD HTML 4.01 Transitional//EN"
"http://www.w3.org/TR/html4/loose.dtd">
<html>
<head>
<meta http-equiv="Content-Type" content="text/html; charset=gb2312">
<title>背景自动变色</title>
</head>
<body>
背景自动变色
<script language="javascript">
var Arraycolor=new
Array("#00FF66","#FFFF99","#99CCFF","#FFCCFF","#FFCC99","#00FFFF","#FFFF00","#FFCC00","#FF00FF");
    var n=0;
    function turncolors(){
        n++;
        if (n==(Arraycolor.length-1)) n=0;
        document.bgColor = Arraycolor[n];
        setTimeout("turncolors()",1000);
    }
    turncolors();
</script>
```

```
    </body>
</html>
```

任务三 编写选项卡效果网页

编写一个网页文件,实现选项卡效果,点击不同选项卡时,可以切换显示内容。

```
<!DOCTYPE html PUBLIC "-//W3C//DTD XHTML 1.0 Transitional//EN" "http://www.w3.org/TR/xhtml1/DTD/xhtml1-transitional.dtd">
<html xmlns="http://www.w3.org/1999/xhtml">
<head>
<title>点击切换选项卡代码</title>
<style type="text/css">
*{margin:0;padding:0;list-style-type:none;}
a,img{border:0;}
body{font:12px/180% Arial, Helvetica, sans-serif,"新宋体";}
.tab1{width:401px;border-top:#cccccc solid 1px;border-bottom:#cccccc solid 1px;margin:50px auto 0 auto;}
.menu{height:28px;border-right:#cccccc solid 1px;}
.menu li{float:left;width:99px;text-align:center;line-height:28px;height:28px;cursor:pointer;border-left:#cccccc solid 1px;color:#666;font-size:14px;overflow:hidden;background:#E0E2EB;}
.menu li.off{background:#FFFFFF;color:#336699;font-weight:bold;}
.menudiv{height:200px;border-left:#cccccc solid 1px;border-right:#cccccc solid 1px;border-top:0;background:#fefefe}
.menudiv div{padding:15px;line-height:28px;}
</style>
<script type="text/javascript">
function setTab(name,cursel){
    cursel_0 = cursel;
    for(var i=1; i<=links_len; i++){
        var menu = document.getElementById(name+i);
        var menudiv = document.getElementById("con_"+name+"_"+i);
        if(i==cursel){
            menu.className="off";
            menudiv.style.display="block";
        }
        else{
            menu.className="";
            menudiv.style.display="none";
        }
    }
}
function Next(){
    cursel_0++;
    if (cursel_0>links_len)cursel_0=1
    setTab(name_0,cursel_0);
```

```
        }
    var name_0 = 'one';
    var cursel_0 = 1;
    var ScrollTime = 3000;//循环周期(毫秒)
    var links_len,iIntervalId;
    onload = function(){
        var links = document.getElementById("tab1").getElementsByTagName('li');
        links_len = links.length;
        for(var i=0; i<links_len; i++){
            links[i].onmouseover = function(){
                clearInterval(iIntervalId);
                this.onmouseout = function(){
                    iIntervalId = setInterval(Next,ScrollTime);;
                }
            }
        }
        document.getElementById("con_" + name_0 + "_" + links_len).parentNode.onmouseover = function(){
            clearInterval(iIntervalId);
            this.onmouseout = function(){
                iIntervalId = setInterval(Next,ScrollTime);;
            }
        }
        setTab(name_0,cursel_0);
        iIntervalId = setInterval(Next,ScrollTime);
    }
</script>
</head>
<body>
<div class="tab1" id="tab1">
    <div class="menu">
        <ul>
            <li id="one1" onclick="setTab('one',1)">首页</li>
            <li id="one2" onclick="setTab('one',2)">点击看看</li>
            <li id="one3" onclick="setTab('one',3)">会自动的</li>
            <li id="one4" onclick="setTab('one',4)">我的网站</li>
        </ul>
    </div>
    <div class="menudiv">
        <div id="con_one_1">我的网站</div>
        <div id="con_one_2" style="display:none;">JS代码,导航菜单</div>
        <div id="con_one_3" style="display:none;">看到效果了吗???</div>
        <div id="con_one_4" style="display:none;">我的网站我做主</div>
    </div>
</div>
<div style="text-align:center;clear:both;"></div>
</body>
</html>
```

代码运行效果见图 2.2。

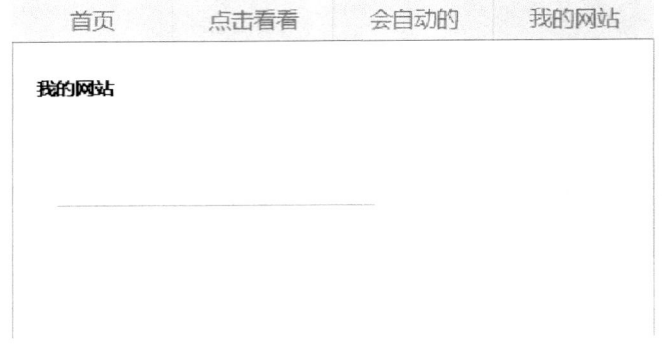

图 2.2 选项卡

任务四 日 期 计 算

编写一个网页文件，JavaScript 计算某一天是星期几，文本框中是默认值，只要按此种格式输入日期时间，就可以推算出当天是星期几。

```
<html>
<head>
<title>计算某一天是星期几</title>
<style type="text/css">
.style5 {font-size: 12px}
</style>
</head>
<script language="javascript">
function checktext()
{
    if((form1.yeartext.value == "") && (form1.monthtext.value == "") && (form1.datetext.value == ""))
    {
        alert("请输入相关信息!");
        form1.yeartext.focus();return;
    }
    if((form1.yeartext.value.length != 4) && (form1.monthtext.value.length != 1) && (form1.datetext.value.length != 1))
    {
        alert("输入错误,只能输入4位数!");
        form1.yeartext.focus();return;
    }
}
function mod(x, x_div)
{
    for (var i=x; i>=x_div; i-=x_div);
        return i;
}
function getday()
{
```

```javascript
var currentyear = parseInt(form1.yeartext.value,10);
var currentmonth = parseInt(form1.monthtext.value,10);
var currentday = parseInt(form1.datetext.value,10);
var sig_val;
var begindate = new Array(0,3,3,6,1,4,6,2,5,0,3,5);
var rundate = new Array(-1,2,2,5,0,3,5,1,4,-1,2,4);
var Pmonth = new Array(29,31,28,31,30,31,30,31,31,30,31,30,31);
var montharray = new Array("星期日","星期一","星期二","星期三","星期四","星期五","星期六");
sig_val = begindate[currentmonth-1];
var val1 = mod((currentyear + parseInt(currentyear/4) + currentday + sig_val)-2,7);
var M=parseInt(document.all.monthtext.value);
var D=parseInt(document.all.datetext.value);
if ((currentyear%4==0 && currentyear%100!=0)||(currentyear%400==0))
{
    if ((M<13)&&(M>0)){
        if ((M==2)&&(D>Pmonth[0])){alert('输入错误');document.all.resulttext.value='';}
        else{
            if ((D>Pmonth[M])&&(M!=2)){alert('输入错误');document.all.resulttext.value='';}
            else{
                sig_val = rundate[currentmonth-1];
                val1 = mod((currentyear + parseInt(currentyear/4) + currentday + sig_val)-2,7);
                if (M>2){val1+=1;}
                form1.resulttext.value = montharray[val1];
            }
        }
    }else{alert('输入错误');document.all.resulttext.value='';}
}
else
{
    if ((M<13)&&(M>0)){
        if (D>Pmonth[M]){alert('输入错误');document.all.resulttext.value='';}
        else{form1.resulttext.value = montharray[val1];}
    }else{alert('输入错误');document.all.resulttext.value='';}
}
}
</script>
<body>
<center>
<form name="form1" method="post" action="">
    <table width="308" border="1" cellpadding="3" cellspacing="1" bordercolor="#33CCFF" bgcolor="#CCFFFF">
        <tr bgcolor="#FFFFFF">
            <td align="center" class="style5">输入年:</td>
            <td width="170"><input name="yeartext" type="text" id="yeartext" value="2016"></td>
        </tr>
        <tr bgcolor="#FFFFFF">
```

```html
            <td align="center" class="style5">输入月:</td>
            <td><input name="monthtext" type="text" value="2"></td>
        </tr>
        <tr bgcolor="#FFFFFF">
            <td align="center" class="style5">输入日:</td>
            <td><input name="datetext" type="text" value="2"></td>
        </tr>
        <tr bgcolor="#FFFFFF">
            <td align="center"><span class="style5">星  期:</span></td>
            <td><input name="resulttext" type="text" id="resulttext"></td>
        </tr>
        <tr align="center" bgcolor="#FFFFFF">
            <td colspan="2">
              <div align="right">
                <input name="enter" type="button" value="计算" onClick="checktext();getday();">
              </div>
            </td>
        </tr>
      </table>
    </form>
  </center>
 </body>
</html>
```

代码运行效果见图 2.3。

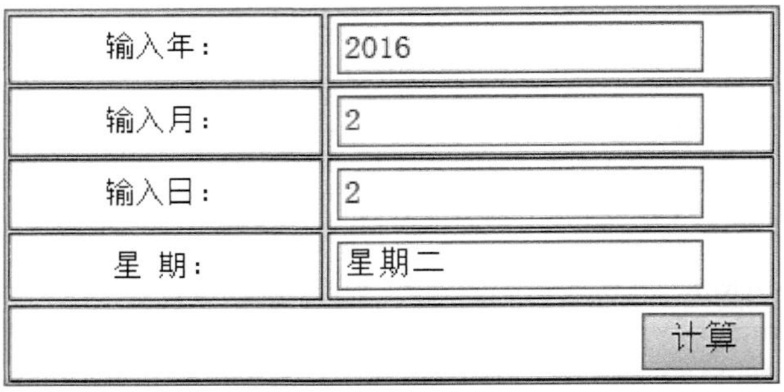

图 2.3　日期计算

任务五　编写自动隐藏显示网页

编写一个网页文件,JavaScript 定时隐藏/显示图片,设定图片在几秒后会自动显示,也会自动隐藏。

```html
<!DOCTYPE HTML PUBLIC "-//W3C//DTD HTML 4.01 Transitional//EN" "http://www.w3.org/TR/html4/loose.dtd">
<html>
<head>
<title>定时隐藏图片</title>
</head>
```

```
<SCRIPT LANGUAGE="JavaScript">
var sec=10;
var timer;
function hidepic()
{
    sec--;
    if(sec==0){
        textfield.value = "图片被隐藏";
        soccer.style.visibility =(soccer.style.visibility == "hidden") ? "visible" : "hidden";
    }
    else{
        textfield.value = "图片会在"+sec+"秒后隐藏";
        setTimeout("hidepic()",1000);
    }
}
</SCRIPT>
<body onLoad = "hidepic();">
<center>
    <input name="textfield" type="text" size="20"><br>
    <DIV ID="soccer"style="position:absolute; left:333px; top:43px">
    <img border="0" src="http://www.codefans.net/jscss/demoimg/wall_s9.jpg">
    </DIV>
</center>
</body>
</html>
```

代码运行效果见图2.4。

图2.4 自动隐藏显示

任务六 文字旋转效果

编写一个网页文件,将文字排成空心圆形呈3D旋转状,实现的3D旋转。

```
<html>
<head>
<meta http-equiv="Content-Type" content="text/html; charset=gb2312">
<title>文字的旋转</title>
<script language="javascript">
Phrase="欢迎光临源码爱好者网页特效";
Balises="";
Taille=40;
```

```
    Midx = 150;
    Decal = 0.5;
    Nb = Phrase.length;
    y = -10000;
    for(x=0;x<Nb;x++){
        Balises = Balises + '<DIV id=L'+x+' STYLE="width:3;font-family:Courier New;font-weight:bold;position:absolute;top:160;left:70;z-index:0">'+Phrase.charAt(x)+'</DIV>'
    }
    document.write(Balises);
    Time = window.setInterval("Alors()",50);
    Alpha = 5;
    I_Alpha = 0.05;
    function Alors(){
        Alpha = Alpha - I_Alpha;
        for (x=0;x<Nb;x++){
            Alpha1 = Alpha + Decal * x;
            Cosine = Math.cos(Alpha1);
            Ob = document.all("L"+x);
            Ob.style.posLeft = Midx + 100 * Math.sin(Alpha1);
            Ob.style.zIndex = 20 * Cosine;
            Ob.style.fontSize = Taille + 30 * Cosine;
            Ob.style.color = "rgb("+(27+Cosine*80+50)+","+(127+Cosine*80+50)+",0)";
        }
    }
</script>
</head>
<body>
</body>
</html>
```

代码运行效果见图 2.5。

图 2.5 文字旋转

五、考核标准

（1）版面布局合理清晰，整体效果美观，观赏性强。（20 分）

（2）网页中没有明显的错误（如超链接、图片无法显示、错别字等）。（30 分）

（3）功能齐全、符合要求。（40 分）

（4）其他效果。（10 分）

项目三　构建动态 Web 开发环境

一、实训目的

- 掌握 Apache 服务器的安装方法；
- 掌握 PHP 服务器的配置方法。

二、实训要求

- 下载安装包，安装 Apache 服务器；
- 下载安装包，配置 PHP 服务器。

三、实训设计

PHP 开发要使用的服务器是 Apache，它是世界使用排名第一的 Web 服务器软件。它可以运行在几乎所有广泛使用的计算机平台上，由于其跨平台和安全性被广泛使用，是最流行的 Web 服务器端软件之一。

安装之前先准备好软件：

Apache 安装包：Apache_2.0.55-win32-x86-no_ssl.msi；

PHP 安装包：PHP-5.0.5-Win32.zip。

四、实训内容

任务一　安装 Apache，配置网站服务器

（1）运行下载好的"Apache_2.0.55-win32-x86-no_ssl.msi"，出现 Apache HTTP Server 2.0.55 的安装向导界面，点"Next"继续，见图 3.1。

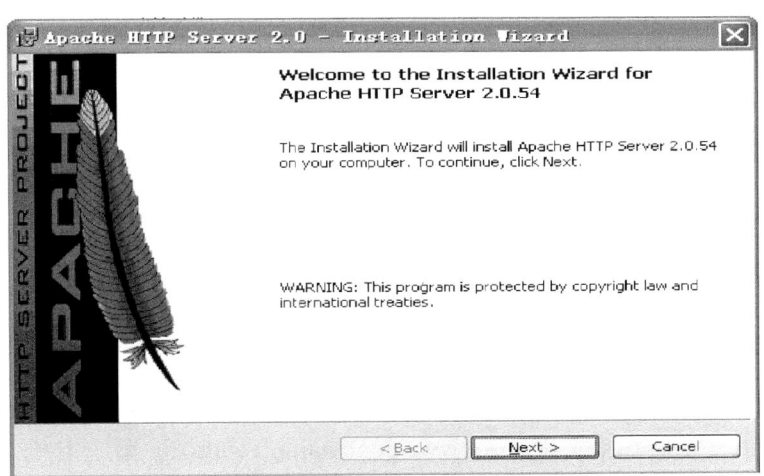

图 3.1　安装向导界面

(2) 确认同意软件安装使用许可条例,选择"I accept the terms in the license agreement",点"Next"继续,见图 3.2。

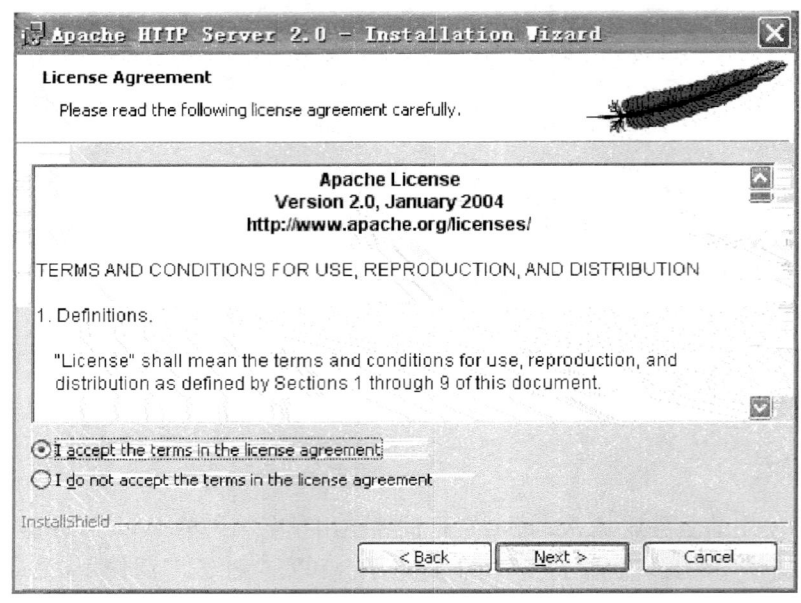

图 3.2　许可协议

(3) 将 Apache 安装到 Windows 上的使用须知,请阅读完毕后,按"Next"继续,见图 3.3。

图 3.3　使用须知

(4) 设置系统信息,在 Network Domain 下填入您的域名(比如:goodwaiter.com),在 Server Name 下填入您的服务器名称(比如:www.goodwaiter.com,也就是主机名加上域名),在 Administrator's Email Address 下填入系统管理员的联系电子邮件地址(比如:yinpeng@

xinhuanet.com），上述三条信息仅供参考，其中联系电子邮件地址会在当系统故障时提供给访问者，三条信息均可任意填写，无效的也行。下面有两个选择，图片上选择的是为系统所有用户安装，使用默认的 80 端口，并作为系统服务自动启动；另外一个是仅为当前用户安装，使用端口 8080，手动启动。一般选择如图所示。按"Next"继续，见图 3.4。

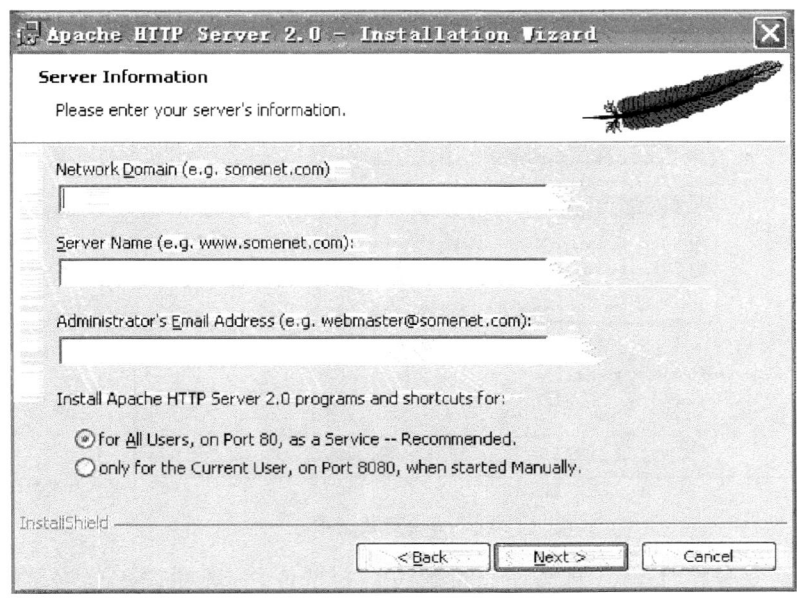

图 3.4　用户信息

（5）选择安装类型，Typical 为默认安装，Custom 为用户自定义安装，我们这里选择 Custom，有更多可选项。按"Next"继续，见图 3.5。

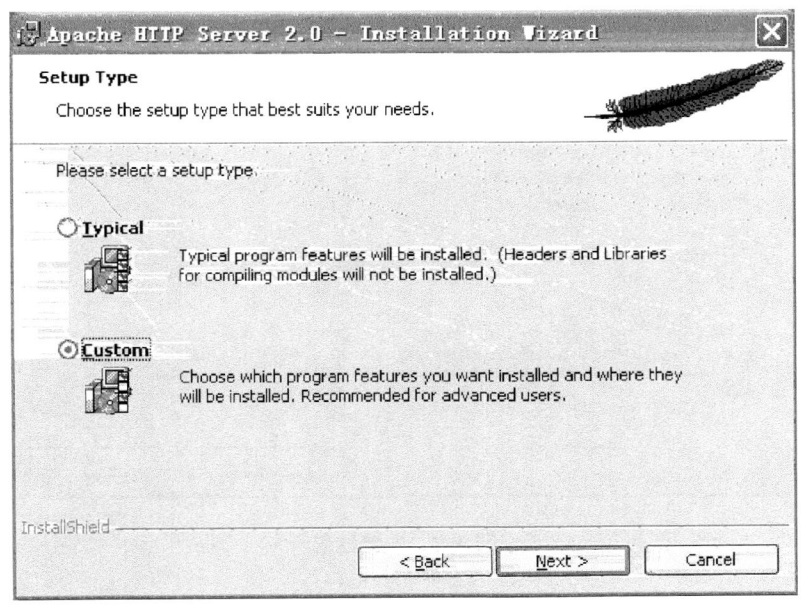

图 3.5　选择安装类型

（6）出现选择安装选项界面，见图3.6，左键点选"Apache HTTP Server 2.0.55"，选择"This feature, and all subfeatures, will be installed on local hard drive."，即"此部分，及下属子部分内容，全部安装在本地硬盘上"。点选"Change..."，手动指定安装目录。

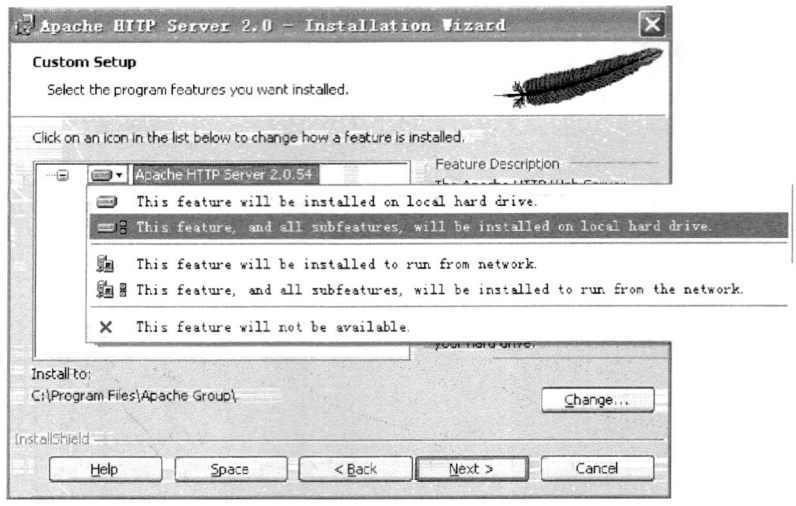

图3.6　安装选项

（7）这里选择安装在"D:\"，具体安装位置可自行选取，一般建议不要安装在操作系统所在盘，免得操作系统坏了之后，还原操作把Apache配置文件也清除了。选"OK"继续，见图3.7。

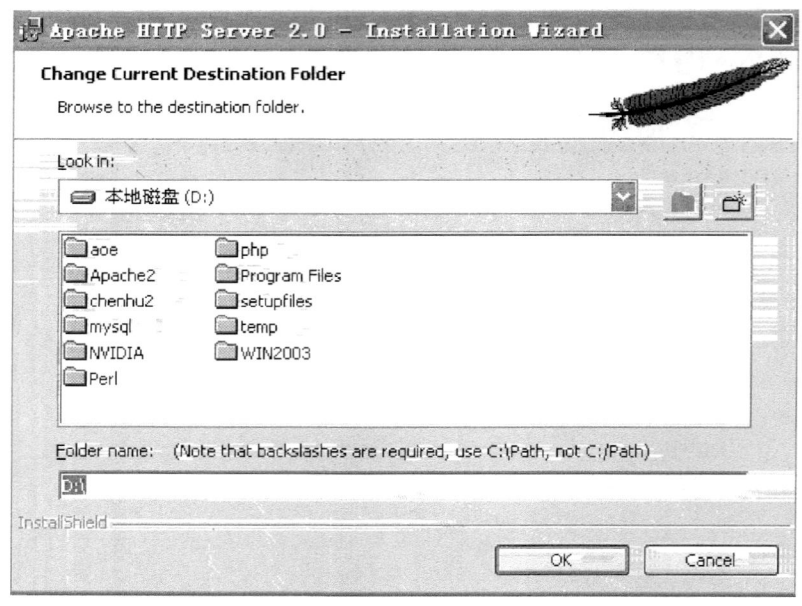

图3.7　安装位置

（8）返回刚才的界面，选"Next"继续，见图3.8。

（9）确认安装选项无误，如果您认为要再检查一遍，可以点"Back"一步步返回检查。点"Install"开始按前面设定的安装选项安装，见图3.9。

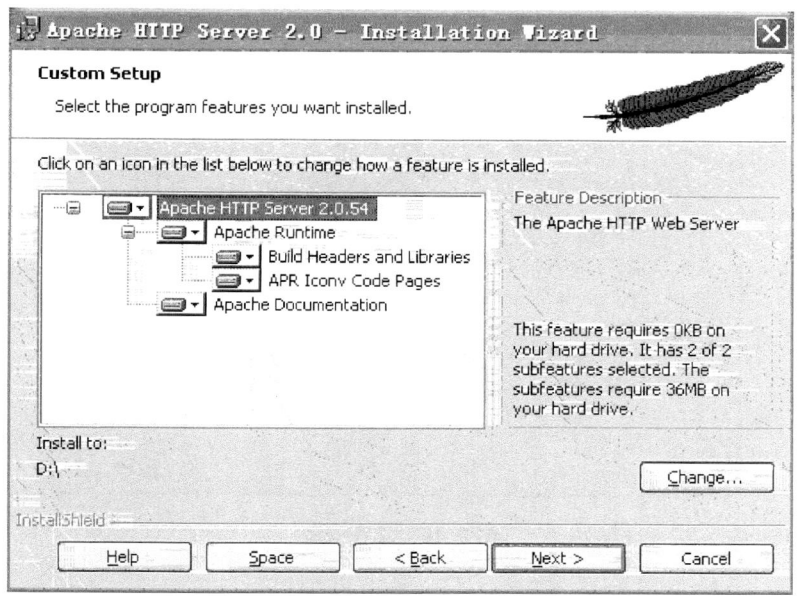

图 3.8　安装选择

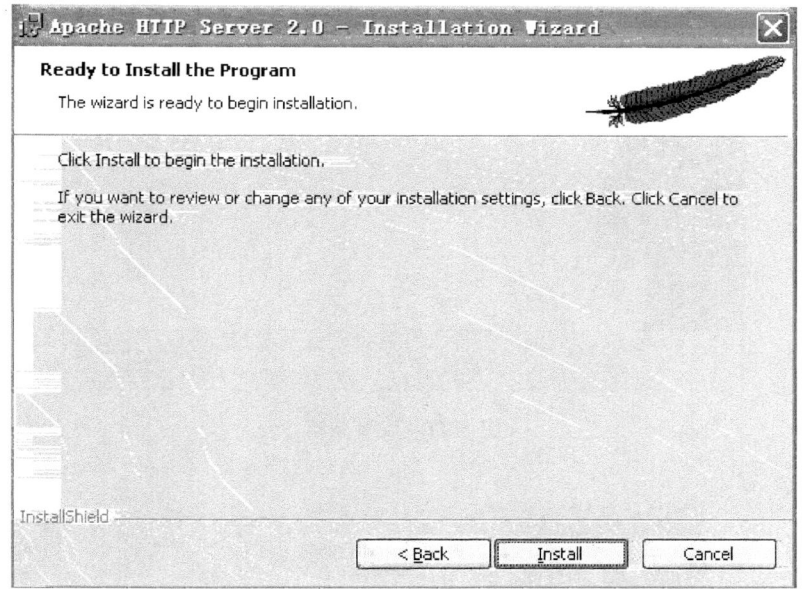

图 3.9　开始安装

（10）如图 3.10 所示，正在安装界面，请耐心等待。

（11）安装向导成功完成，按"Finish"结束 Apache 的软件安装，见图 3.11。

（12）右下角状态栏出现了一个绿色图标，表示 Apache 服务已经开始运行，见图 3.12。

（13）我们来熟悉一下这个图标，很方便的，在图标上左键单击，出现如下界面，有"Start（启动）""Stop（停止）""Restart（重启动）"三个选项，可以很方便地对安装的 Apache 服务器进行上述操作，见图 3.13。

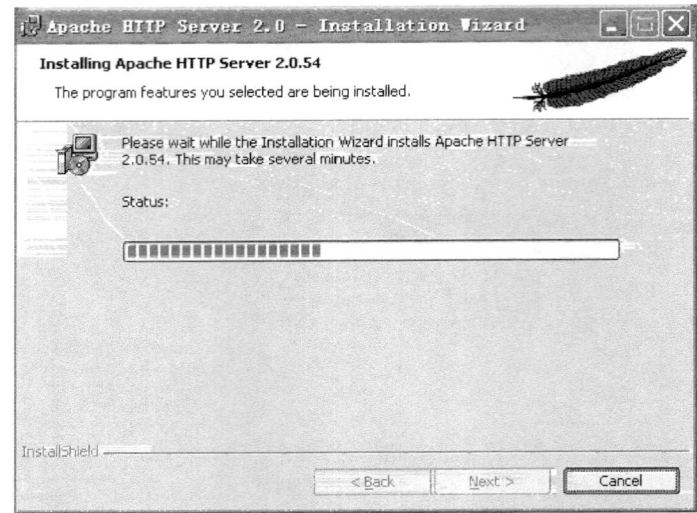

图 3.10　正在安装

图 3.11　安装完毕

图 3.12　运行图标　　　　　　　　图 3.13　图标选项

（14）测试一下按默认配置运行的网站界面，在 IE 地址栏打"http://127.0.0.1"，点"转到"，就可以看到图 3.14 的页面，表示 Apache 服务器已安装成功。

（15）现在开始配置 Apache 服务器，使它更好地为我们服务，事实上，如果不配置，你的安装目录下的 Apache2\htdocs 文件夹就是网站的默认根目录，在里面放入文件就可以了。这里我们还是要配置一下，有什么问题或修改，配置始终是要会的，如图 3.15 所示，"开始""所有程序"

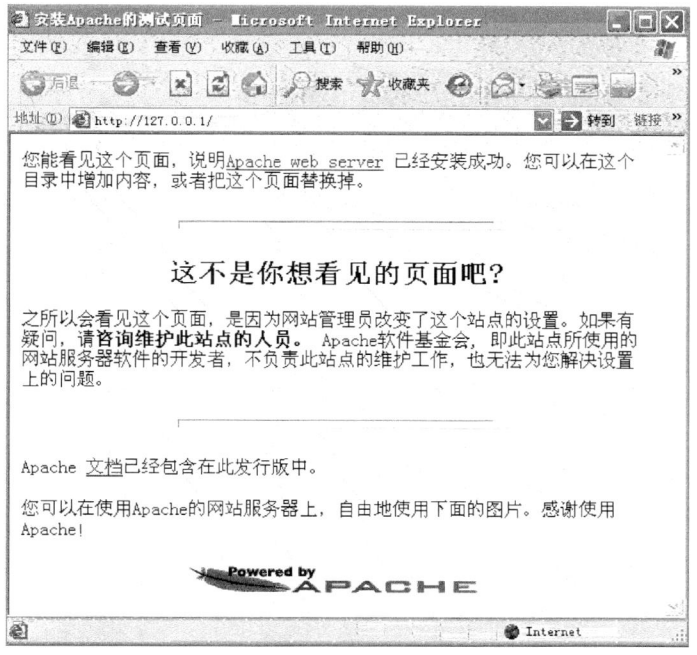

图 3.14　首页

"Apache HTTP Server 2.0.55""Configure Apache Server""Edit the Apache httpd conf Configuration file",点击"打开",见图 3.15。

图 3.15　配置服务器

（16）XP 的记事本有了些小变化,很实用的一个功能就是可以看到文件内容的行、列位置,按图 3.16 所示,点"查看",勾选"状态栏",界面右下角就多了个标记,"Ln 78, Col 10"就表示"行 78,列 10",这样可以迅速地在文件中定位,方便解说。当然,你也可以通过"编辑","查找"输入关键字来快速定位。每次配置文件的改变,保存后,必须在 Apache 服务器重启动后生效,可以用前面讲的小图标方便地控制服务器随时"重启",见图 3.16。

（17）现在正式开始配置 Apache 服务器,"Ln 228",或者查找关键字"DocumentRoot"（也就是网站根目录）,找到如图 3.17 所示地方,然后将""内的地址改成你的网站根目录,地址格式请照图上的写,主要是一般文件地址的"\"在 Apache 里要改成"/",见图 3.17。

（18）"Ln 253",同样,你也可以通过查找"<

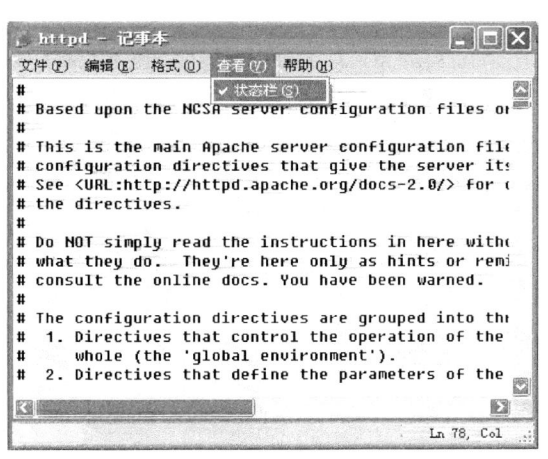

图 3.16　修改配置文件

项目三　构建动态 Web 开发环境／21

DIRECTORY>"来定位,以后不再说明,将""内的地址改成跟 DOCUMENTROOT 的一样,见图 3.18。

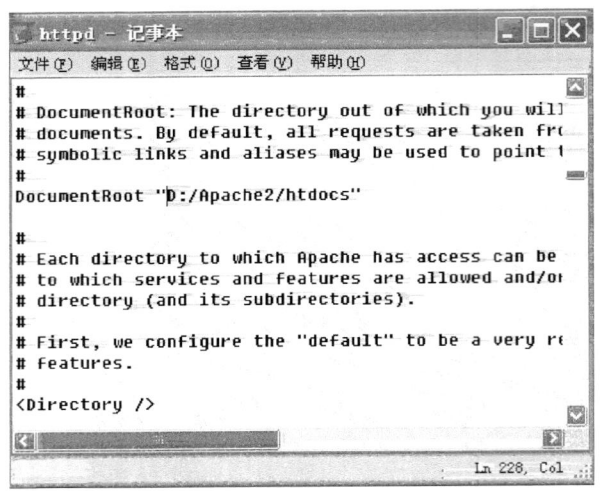

图 3.17 修改 httpd 文件

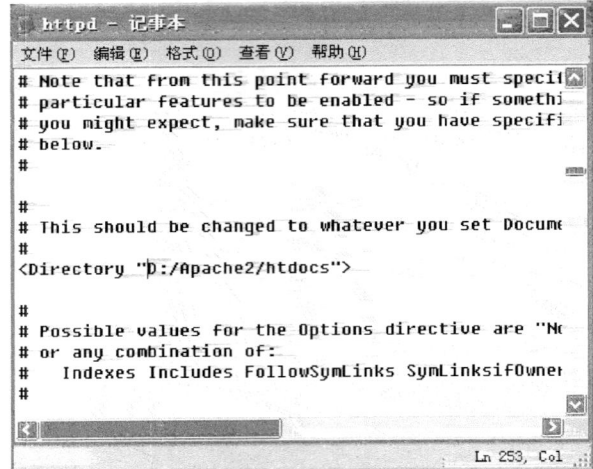

图 3.18 修改 DIRECTORY

（19）"Ln 321",DirectoryIndex(目录索引,也就是在仅指定目录的情况下,默认显示的文件名),可以添加很多,系统会根据从左至右的顺序来优先显示,以单个半角空格隔开,比如有些网站的首页是 index.htm,就在光标那里加上"index.htm"文件名是任意的,不一定非得"index.html",比如"test.php"等,都可以,见图 3.19。

（20）这里有一个选择配置选项,以前可能要配置,现在不用配置了,就是强制所有输出文件的语言编码,html 文件里有语言标记（这个就是设定文档语言为 gb2312）的也会强制转换。如果打开的网页出现乱码,请先检查网页内有没有上述 html 语言标记,如果没有,添加上去就能正常显示了。把"# DefaultLanguage nl"前面的"#"去掉,把"nl"改成你要强制输出的语言,中文是"zh-cn",保存,关闭,见图 3.20。

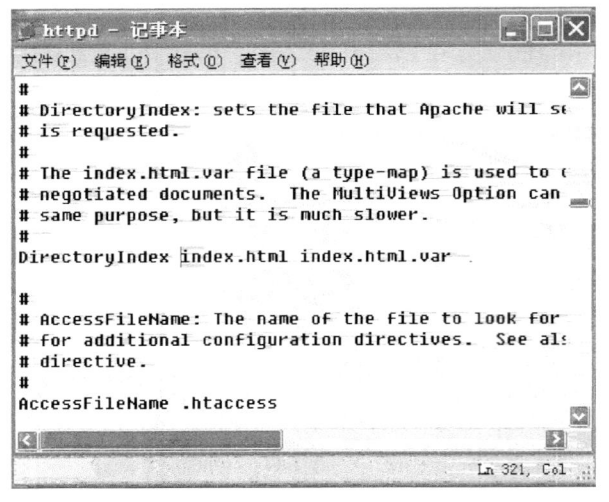

图 3.19 修改目录索引

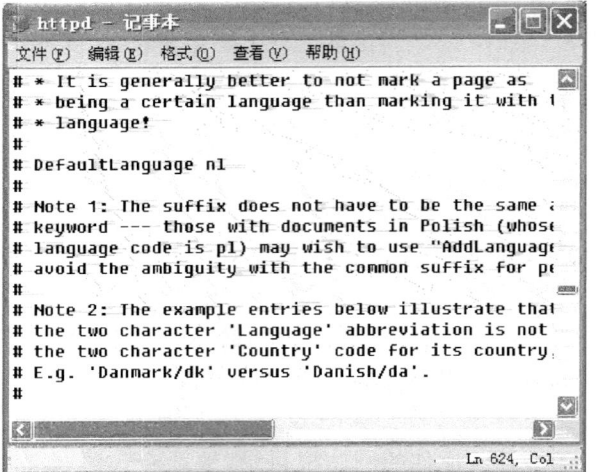

图 3.20 修改选择配置

简单的 Apache 配置就到此结束了,现在利用先前的小图标重启动,所有的配置就生效了,你的网站就成了一个网站服务器,如果你加载了防火墙,请打开 80 或 8080 端口,或者允许 Apache 程序访问网络,否则别人不能访问。

任务二　PHP 的安装与 Apache 结合使用

（1）将下载的 php 安装文件 php-5.0.5-Win32.zip 右键解压缩,见图 3.21。

（2）指定解压缩的位置,设定在"D:\php",见图 3.22。

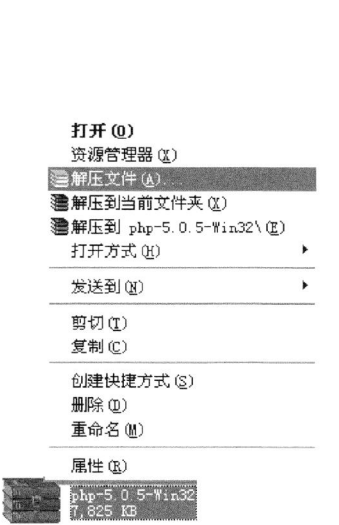

图 3.21　解压做

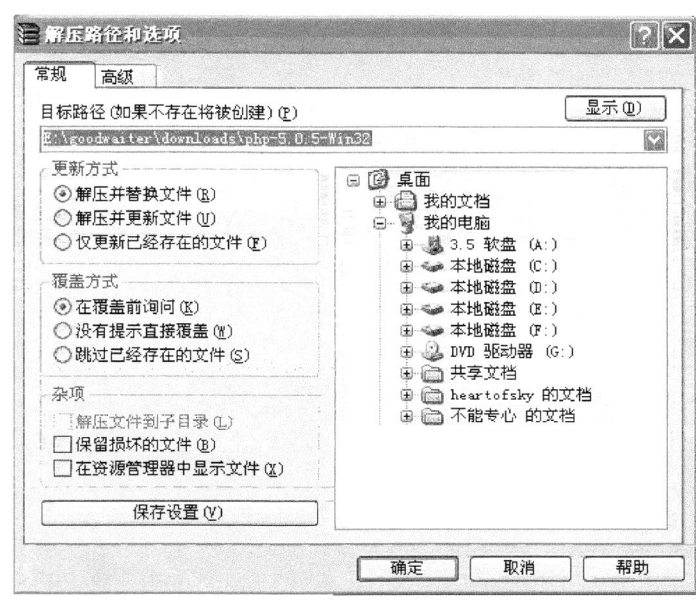

图 3.22　选择解压缩位置

（3）查看解压缩后的文件夹内容,找到"php.ini-dist"文件,将其重命名为"php.ini",打开编辑,找到下面图中的地方,Ln 385,有一个"register_globals = Off"值,这个值是用来打开全局变量的。比如表单送过来的值,如果这个值设为"Off",就只能用"$_POST['变量名']、$_GET['变量名']"等来取得送过来的值；如果设为"On",就可以直接使用"$变量名"来获取送过来的值。当然,设为"Off"就比较安全,不会让人轻易将网页间传送的数据截取。这个值是否改成"On"根据情况选择。见图 3.23。

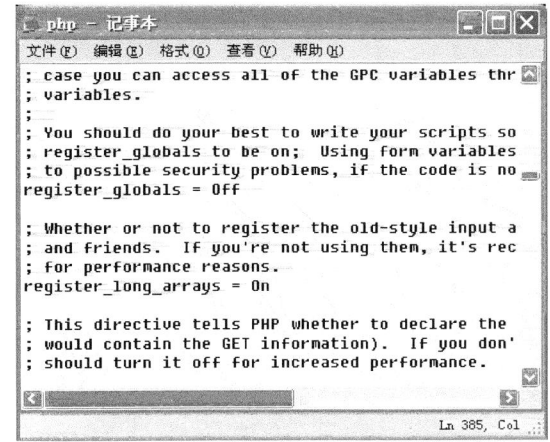

图 3.23　打开 php.ini

（4）这里还有一个地方要编辑,功能就是使 PHP 能够直接调用其他模块,比如访问 mysql,如图 3.24 所示,Ln563,选择要加载的模块,去掉前面的";",就表示要加载此模块了,加载得越多,占用的资源也就多一点,不过也多不到哪去,比如要用 mysql,就要把";extension= php_mysql.dll"前的";"去掉。所有的模块文件都放在 php 解压缩目录的"ext"之下,这里的截图是把所有能加载的模块都加载上去了,前面的";"没去掉的,是因为"ext"目录下默认没有此模块,加载会提示

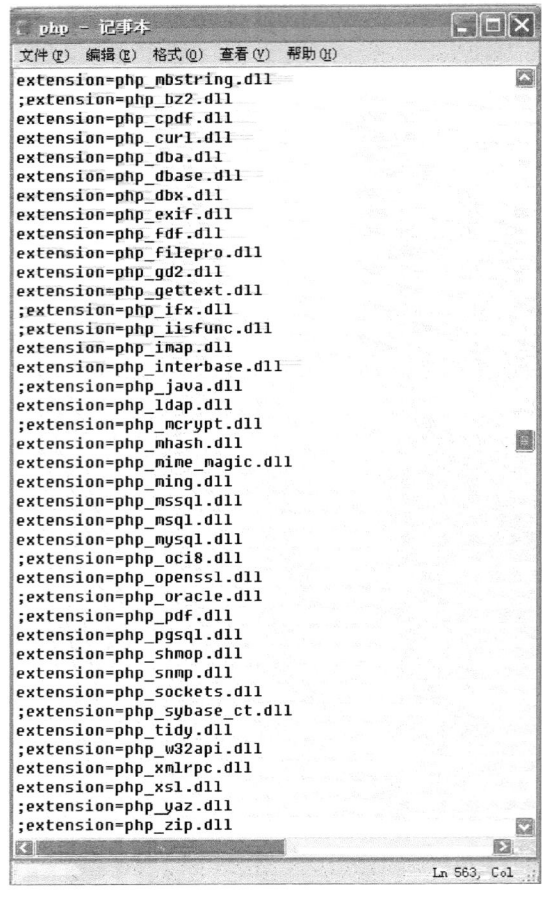

图 3.24 修改 php.ini

找不到文件而出错。这里只是参考，一般不需要加载这么多，需要的加载上就可以了，编辑好后保存，关闭，见图 3.24。

（5）如果上一步加载了其他模块，就要指明模块的位置，否则重启 Apache 的时候会提示"找不到指定模块"的错误，这里介绍一种最简单的方法，直接将 php 安装路径、里面的 ext 路径指定到 Windows 系统路径中——在"我的电脑"上右键，"属性"，选择"高级"标签，点选"环境变量"，在"系统变量"下找到"Path"变量，选择，双击或点击"编辑"，将"；D:\php；D:\php\ext"加到原有值的后面，当然，其中的"D:\php"是我的安装目录，你要将它改为自己的 php 安装目录，如图 3.25 所示，全部确定。系统路径添加好后要重启电脑才能生效，可以现在重启，也可以在所有软件安装或配置好后重启，见图 3.25。

（6）在开始将 php 以 module 方式与 Apache 相结合，使 php 融入 Apache，照先前的方法打开 Apache 的配置文件，Ln 173，找到这里，添加进如图 3.26 所示选中的两行，第一行"LoadModule php5_module D:/php/php5apache2.dll"是指以 module 方式加载 php，第二行"PHPIniDir "D:/php""是指明 php 的配置文件 php.ini 的位置，当然，其中的"D:/php"要改成你先前选择的 php 解压缩的目录，见图 3.26。

图 3.25 编辑系统变量

(7) 还是 Apache 的配置文件，Ln 757，加入"AddType application/x-httpd-php．php""AddType application/x-httpd-php．html"两行，你也可以加入更多，实质就是添加可以执行 php 的文件类型，比如你再加上一行"AddType application/x-httpd-php．htm"，则．htm 文件也可以执行 php 程序了，你甚至还可以添加上一行"AddType application/x-httpd-php．txt"，让普通的文本文件格式也能运行 php 程序，见图 3.27。

(8) 前面所说的目录默认索引文件也可以改一下，因为现在加了 php，有些文件就直接存为．php 了，我们也可以把"index．php"设为默认索引文件，优先顺序就自己排了，我的是放在第一位。编辑完成，保存关闭，见图 3.28。

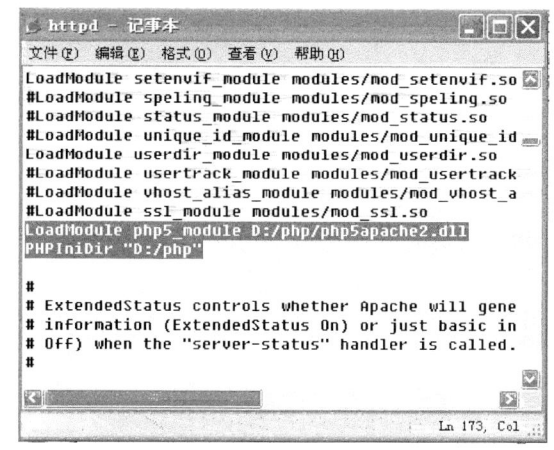

图 3.26　编辑 module

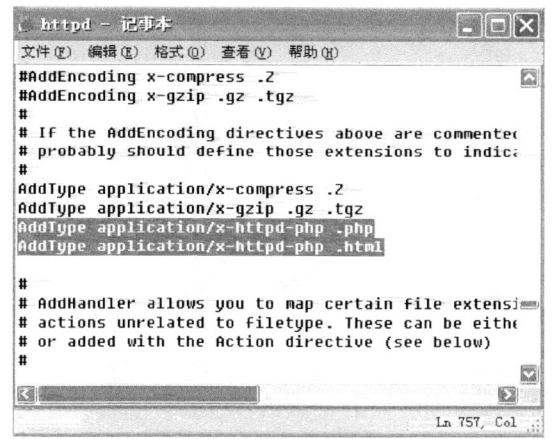

图 3.27　添加可以执行 php 的文件类型

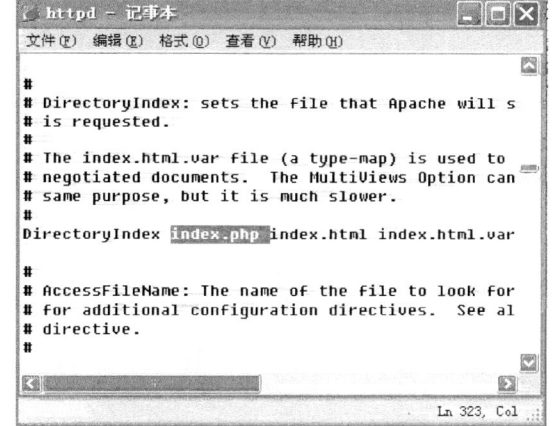

图 3.28　编辑目录默认索引

至此，php 的安装，与 Apache 的结合已经全部完成，用屏幕右下角的小图标重启 Apache，你的 Apache 服务器就支持了 php。

五、考核标准

（1）正确下载安装包。（10 分）

（2）完成 Apache 服务器安装。（30 分）

（3）完成 PHP 服务器安装。（30 分）

（4）完成 PHP 配置。（30 分）

项目四 PHP 内部函数应用

一、实训目的

- 掌握 PHP 内部函数语法结构；
- 掌握 PHP 日期时间函数使用方法；
- 掌握 PHP 正则表达式；
- 掌握 PHP 文件和 FTP 函数。

二、实训要求

- 使用 PHP 日期时间函数描述时间；
- 使用正则表达式校验数据；
- 使用 PHP 文件函数和 FTP 函数操作文件。

三、实训设计

PHP 有很多标准的函数和结构，可以实现不同的功能。本实训通过日期和时间函数、文件函数、PHP 正则表达式、FTP 函数来帮助学生理解 PHP 内部函数的定义和用法。

四、实训内容

任务一 PHP 时间日期函数

使用 PHP 时间日期函数，按照要求输出具体日期和时间。

```
<? php
// 假定今天是:March 10th, 2006, 5:16:18 pm
echo $today = date("F j, Y, g:i a")."<br>";          // 结果是:March 10, 2006, 5:16 pm
echo $today = date("m.d.y")."<br>";                  // 结果是:03.10.06
echo $today = date("j,n,Y")."<br>";                  // 结果是:10,3,2006
echo $today = date("Ymd")."<br>";                    // 结果是:20060310
echo $today = date('h-i-s,j-m-y,it is w Day z ')."<br>";  // 05-16-18, 10-03-06, 1631 1618 6 Fripm01
echo $today = date('\i\t \i\s \t\h\e jS \d\a\y.')."<br>"; //结果是:It is the 10th day.
echo $today = date("D M j G:i:s T Y")."<br>";        //结果是: Sat Mar 10 15:16:08 MST 2006
echo $today = date('H:m:s \m \i\s\ \m\o\n\t\h')."<br>"; // 结果是:17:03:18 m is month
echo $today = date("H:i:s")."<br>";                  // 17:16:18
? >
```

任务二 PHP 文件函数

使用 PHP 文件函数，打开文件，并读取文件内容，最后向文件写入新的内容，关闭文件。

```php
<?php
//复制文件示例
if(copy('./file1.txt','../data/file2.txt')){
echo "文件复制成功!";
}else{
echo "文件复制失败!";
}

//删除文件示例
$filename = "file.txt";
if (file_exists($filename)){
if (unlink($filename)){
echo "文件删除成功!";
}else{
echo "文件删除失败!";
}
}else{
echo "目标文件不存在";
}

//重命名文件示例
if (rename('./demo.php', './demo.html')){
echo "文件重命名成功!";
}else{
echo "文件重命名失败";
}

//截取文件示例
$fp = fopen('./data.txt', "r+") or die('文件打开失败');
if(ftruncate($fp, 1024)){
echo "文件截取成功!";
}else{
echo "文件截取失败!";
}
?>
```

任务三　PHP 正则表达式

使用 PHP 正则表达式校验数据，判断电话号码是否符合要求、将字符串分割为数组。

```
<!DOCTYPE html PUBLIC "-//W3C//DTD XHTML 1.0 Transitional//EN" "http://www.w3.org/TR/xhtml1/DTD/xhtml1-transitional.dtd">
<html xmlns="http://www.w3.org/1999/xhtml">
<head>
<meta http-equiv="Content-Type" content="text/html; charset=gb2312" />
<title>无标题文档</title>
</head>

<body><?php
$date="2010-1234567";
```

```php
if(preg_match("[(^\d{4})-(\d{7}$)]",$date,$res))
{echo $res[0]."<br>".$res[1].$res[2]."<br>";}
else
echo "false";

//$date="2010-6-3";
//if(ereg("([0-9]{4})-([0-9]{1,2})-([0-9]{1,2})",$date,$res))
//{echo $res[1].$res[2].$res[3];}
//else
//echo "false";
echo preg_replace("/0/","1",$date)."<br>";
//if(eregi("[0-9]{4}-[0-9]{1,2}-[0-9]{1,2}",$date,$res))
//{echo "true";}
list($a,$b,$c,$d)=explode(",","33,44,55,66");
echo $a."<br>";
echo $b."<br>";
echo $c."<br>";
echo $d."<br>";
$str="w-e-r-tt";
$spli=explode("-",$str);
for($i=0;$i<count($spli);$i++)
{echo $spli[$i]."<br>";}
if(preg_match("[^[0-9]*$]","11123571m",$res))
{echo "true";}
else
echo "false";
?>
</body>
</html>
```

任务四　FTP　函　数

使用 FTP 函数访问 FTP 服务器,并完成数据的上传和下载。

```php
<?
$ftp_server="192.168.1.127";
    $ftp_user = "foo";
    $ftp_pass = "bar";
//连接 ftp 服务器
    $con=ftp_connect($ftp_server);
//发送用户名和密码
    ftp_login($con,$ftp_user,$ftp_pass);
//取得服务器的系统类型
    ftp_systype($con);
//列出文件
    $filelist=ftp_nlist($con,"/");
    foreach ($contents as $entry) {
        echo $entry, "<br />\n";
}
//下载文件
```

```php
    $local_file = 'local.zip';
    $server_file = 'server.zip';
    $down=ftp_get($con,$local_file,$server_file,FTP_BINARY);
    if(!$down){
        echo "no!";
    }else{
    echo "ok!";
}
    //获得当前路径
    echo ftp_pwd($con);
    echo "<br>";
    //改变路径
    ftp_chdir($con,"somedir");
    echo ftp_pwd($con);
    //返回刚才的目录
    ftp_cdup($con);
echo ftp_pwd($con);
    //创建文件夹,删除的函数为 ftp_rmdir($con,$dir)
    $mkdir=ftp_mkdir($con,"test");
    //上传文件
    ftp_put($con,$destination_file, $source_file,FTP_ASCII);
    //上传一个已经打开的文件到 FTP 服务器
    $file = 'somefile.txt';
    $fp = fopen($file,'r');
    ftp_fput($con, $file, $fp, FTP_ASCII)
    //关闭 ftp 连接
    ftp_close($con);
?>
```

五、考核标准

（1）完成日期时间的调用。（25 分）

（2）完成文件的打开、读取、写入。（25 分）

（3）完成数据的校验。（25 分）

（4）完成 FTP 的访问。（25 分）

项目五 MySQL 数据库的安装配置

一、实训目的

- 掌握 MySQL 的安装方法；
- 掌握 MySQL 数据库的基本操作；
- 掌握 Dreamweaver 下动态站点的配置。

二、实训要求

- 完成 MySQL 的安装；
- 配置 Dreamweaver 下的动态站点,运行 Apache＋PHP＋MySQL 的网站。

三、实训设计

本实训将会完成 Apache＋PHP＋MySQL 组合环境中最后一个软件——MySQL 服务器的安装配置。MySQL 是最流行的关系型数据库管理系统,在 Web 应用方面 MySQL 是最好的 RDBMS（Relational Database Management System,关系数据库管理系统）应用软件之一。完成安装后,将会对项目一中安装的 Dreamweaver 重新配置,以适应数据库的需求。

四、实训内容

任务一 安装配置 MySQL 数据

（1）下载 MySQL 安装文件 MySQL-4.1.14-win32.zip,双击解压缩,运行"setup.exe",出现如下界面,MySQL 安装向导启动,按"Next"继续,见图 5.1。

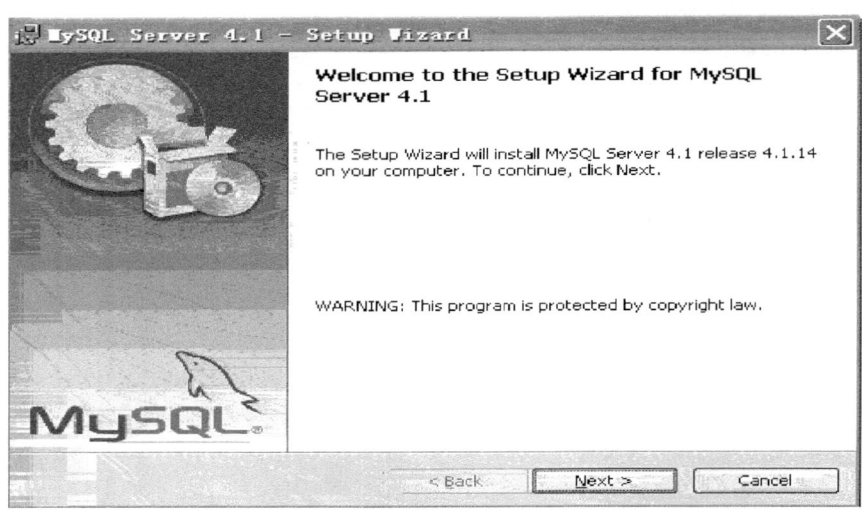

图 5.1 安装向导

（2）选择安装类型，有"Typical（默认）""Complete（完全）""Custom（用户自定义）"三个选项，我们选择"Custom"，有更多的选项，也方便熟悉安装过程，见图 5.2。

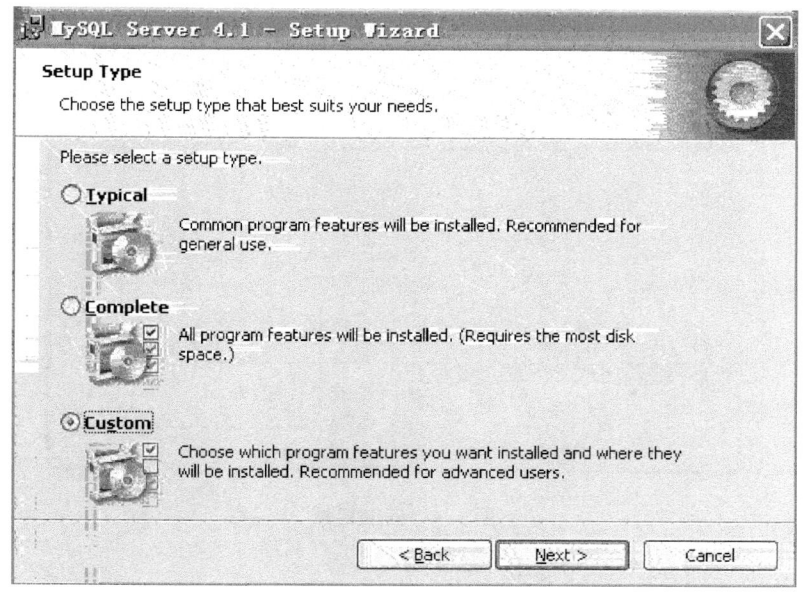

图 5.2　安装类型

（3）"Developer Components（开发者部分）"上左键单击，选择"This feature, and all subfeatures, will be installed on local hard drive."，即"此部分，及下属子部分内容，全部安装在本地硬盘上"。在上面的"MySQL Server（MySQL 服务器）""Client Programs（MySQL 客户端程序）""Documentation（文档）"也如此操作，以保证安装所有文件。点选"Change…"，手动指定安装目录，见图 5.3。

图 5.3　安装文件选择

（4）选择安装目录，我的是"D:\MySQL"，也建议不要放在与操作系统同一分区，这样可以防止系统备份还原的时候，数据被清空。按"OK"继续，见图 5.4。

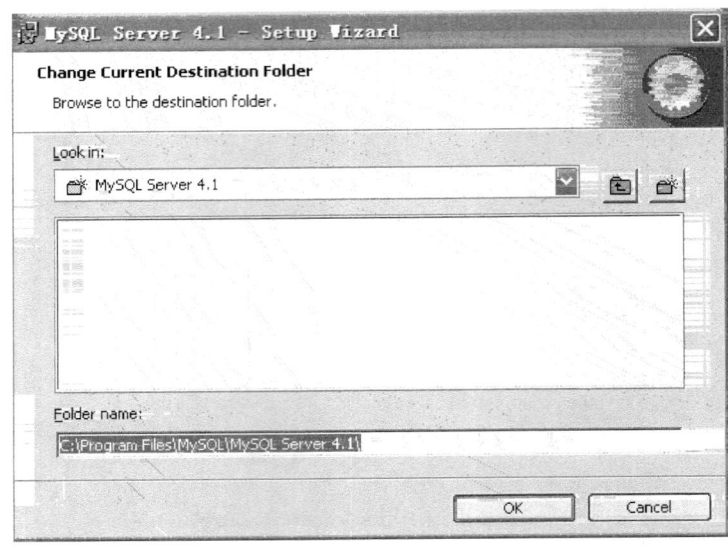

图 5.4　安装目录

(5) 返回刚才的界面，按"Next"继续，见图 5.5。

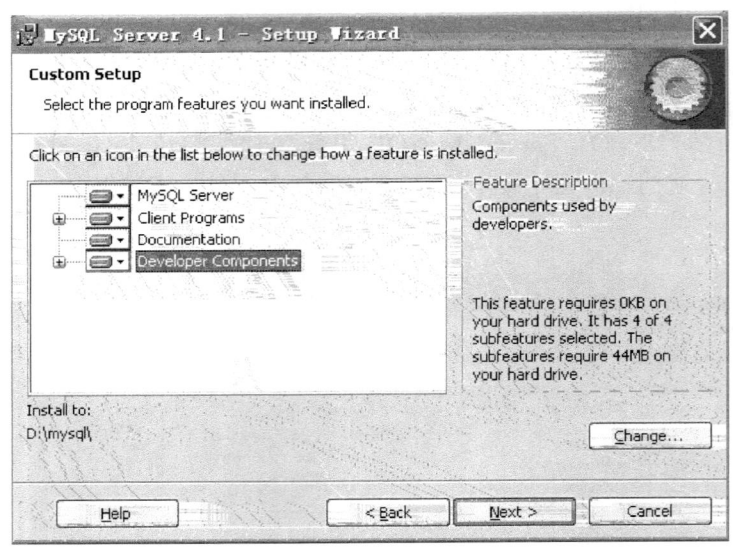

图 5.5　选择完毕

(6) 检查一下先前的设置，如果有误，按"Back"返回重做，按"Install"开始安装，见图 5.6。

(7) 在安装中，请稍候，见图 5.7。

(8) 这里是询问你是否要注册一个 MySQL.com 的账号，或是使用已有的账号登录 MySQL.com，一般不需要了，点选"Skip Sign-Up"，按"Next"略过此步骤，见图 5.8。

(9) 软件安装完成了，出现上面的界面，这里有一个很好的功能，MySQL 配置向导，不用像以前一样，自己手动乱七八糟地配置 my.ini 了，将"Configure the MySQL Server now"前面的勾打上，点"Finish"结束软件的安装并启动 MySQL 配置向导，见图 5.9。

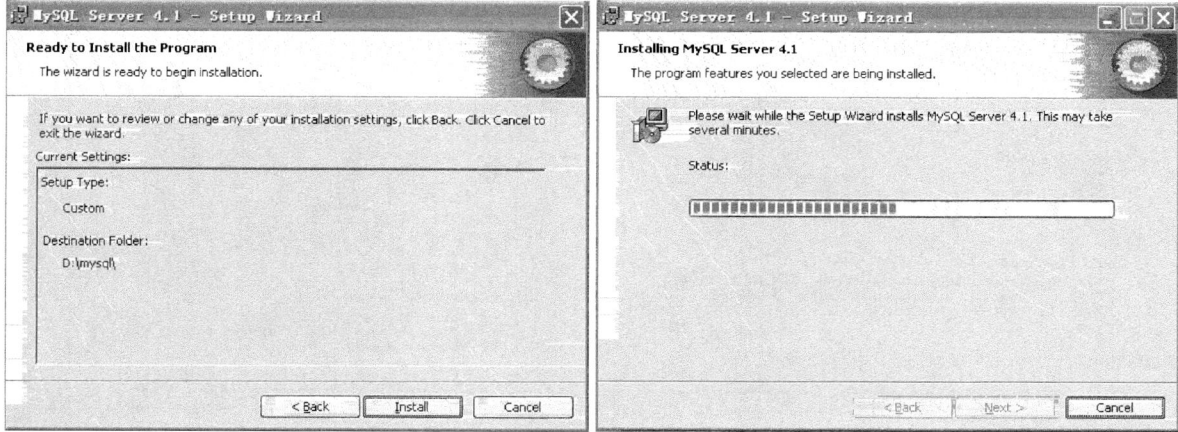

图 5.6　开始安装　　　　　　　　　图 5.7　安装中

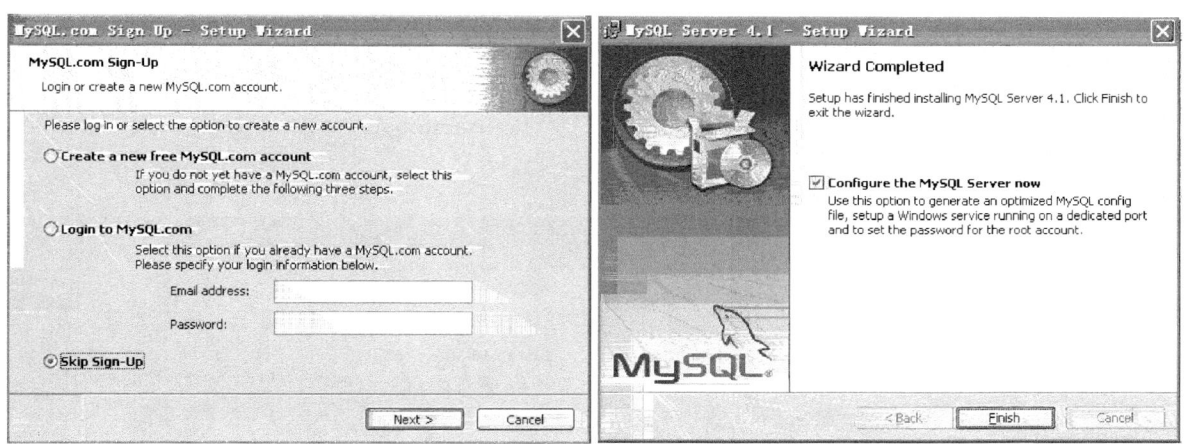

图 5.8　账号密码　　　　　　　　　图 5.9　安装完成

（10）MySQL 配置向导启动界面，按"Next"继续，见图 5.10。

（11）选择配置方式，"Detailed Configuration（手动精确配置）""Standard Configuration（标准配置）"，我们选择"Detailed Configuration"，方便熟悉配置过程，见图 5.11。

（12）选择服务器类型，"Developer Machine（开发测试类，MySQL 占用很少资源）""Server Machine（服务器类型，MySQL 占用较多资源）""Dedicated MySQL Server Machine（专门的数据库服务器，MySQL 占用所有可用资源）"，大家根据自己的类型选择，一般选"Server Machine"，不会太少，也不会占满，见图 5.12。

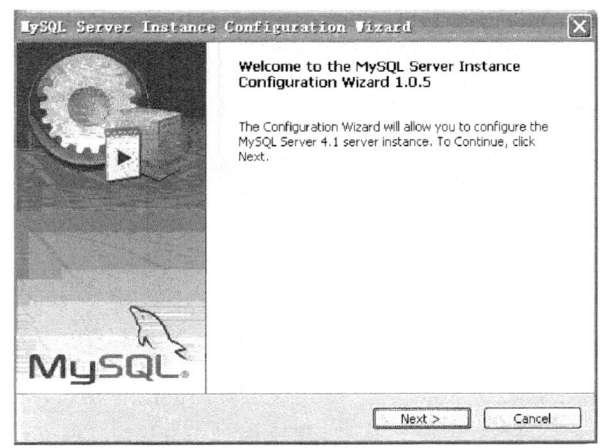

图 5.10　配置向导

项目五　MySQL 数据库的安装配置／**33**

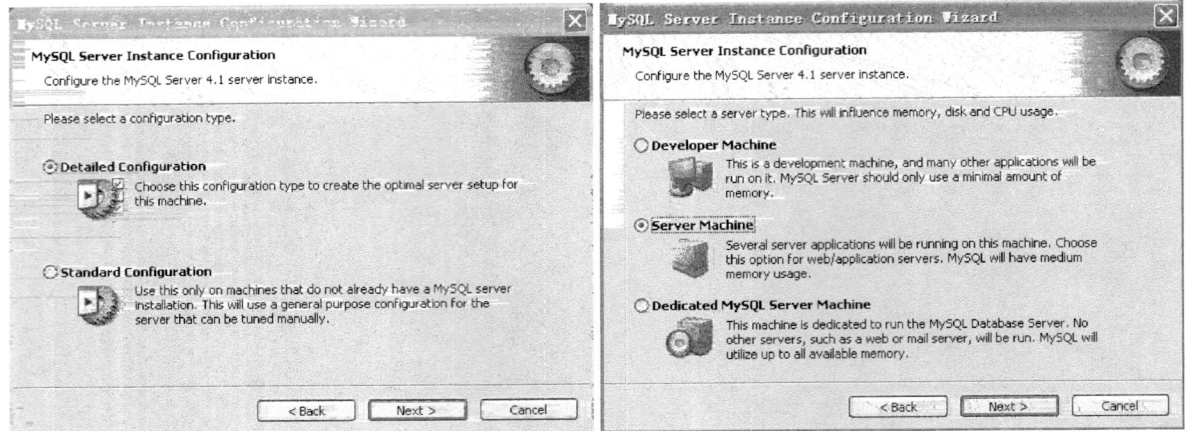

图 5.11　配置方式　　　　　　　　　　　图 5.12　服务器类型

(13) 选择 MySQL 数据库的大致用途,"Multifunctional Database(通用多功能型,好)" "Transactional Database Only(服务器类型,专注于事务处理,一般)""Non-Transactional Database Only(非事务处理型,较简单,主要做一些监控、记费用,对 MyISAM 数据类型的支持仅限于 non-transactional),随自己的用途而选择了,我这里选择"Transactional Database Only",按"Next"继续,见图 5.13。

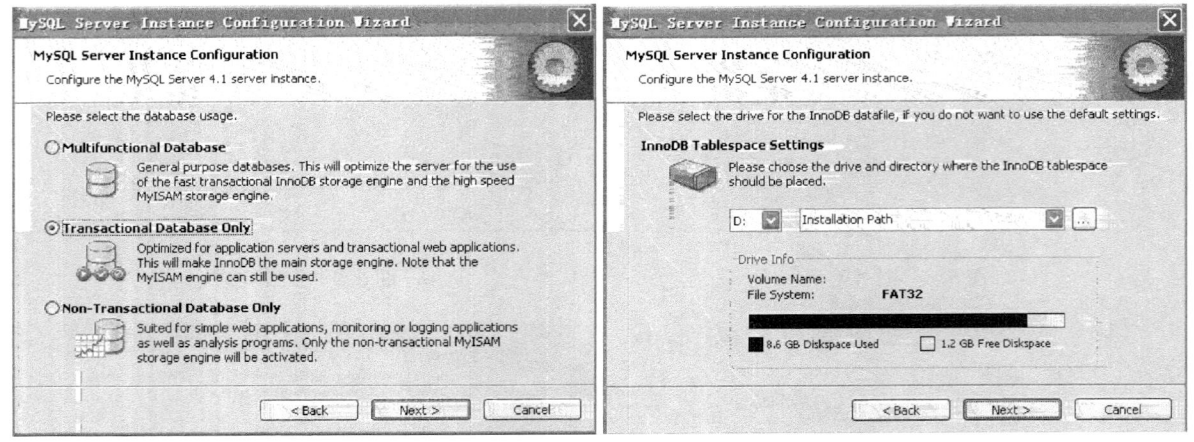

图 5.13　数据库用途　　　　　　　　　　图 5.14　InnoDB Tablespace 选择

(14) InnoDB Tablespace 进行配置,就是为 InnoDB 数据库文件选择一个存储空间,如果修改了,要记住位置,重装的时候要选择一样的地方,否则可能会造成数据库损坏,当然,对数据库做个备份就没问题了,这里不详述。我这里没有修改,使用默认位置,直接按"Next"继续,见图 5.14。

(15) 选择您的网站的一般 MySQL 访问量,同时连接的数目,"Decision Support(DSS)/OLAP(20 个左右)""Online Transaction Processing(OLTP)(500 个左右)""Manual Setting(手动设置,自己输一个数)",我这里选"Online Transaction Processing(OLTP)",自己的服务器,应该够用了,按"Next"继续,见图 5.15。

(16) 是否启用 TCP/IP 连接,设定端口,如果不启用,就只能在自己的机器上访问 MySQL 数据库了,我这里启用,把前面的勾打上,Port Number:3306,按"Next"继续,见图 5.16。

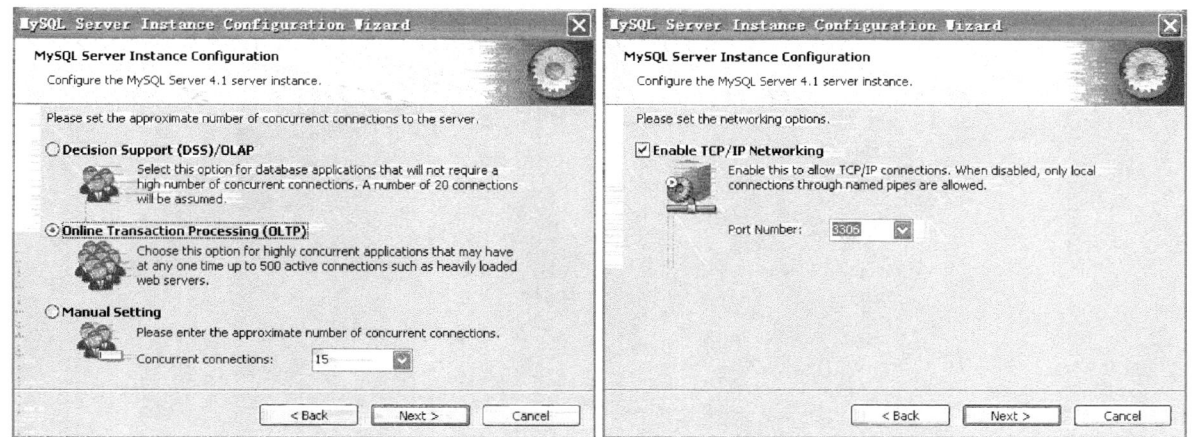

图 5.15　连接数目　　　　　　　　　　　　图 5.16　端口

（17）这个比较重要，就是对 MySQL 默认数据库语言编码进行设置，第一个是西文编码，第二个是多字节的通用 utf8 编码，都不是我们通用的编码，这里选择第三个，然后在 Character Set 那里选择或填入"gbk"，当然也可以用"gb2312"，区别就是 gbk 的字库容量大，包括了 gb2312 的所有汉字，并且加上了繁体字和其他乱七八糟的字——使用 MySQL 的时候，在执行数据操作命令之前运行一次"SET NAMES GBK;"（运行一次就行了，GBK 可以替换为其他值，视这里的设置而定），就可以正常使用汉字（或其他文字）了，否则不能正常显示汉字。按"Next"继续，见图 5.17。

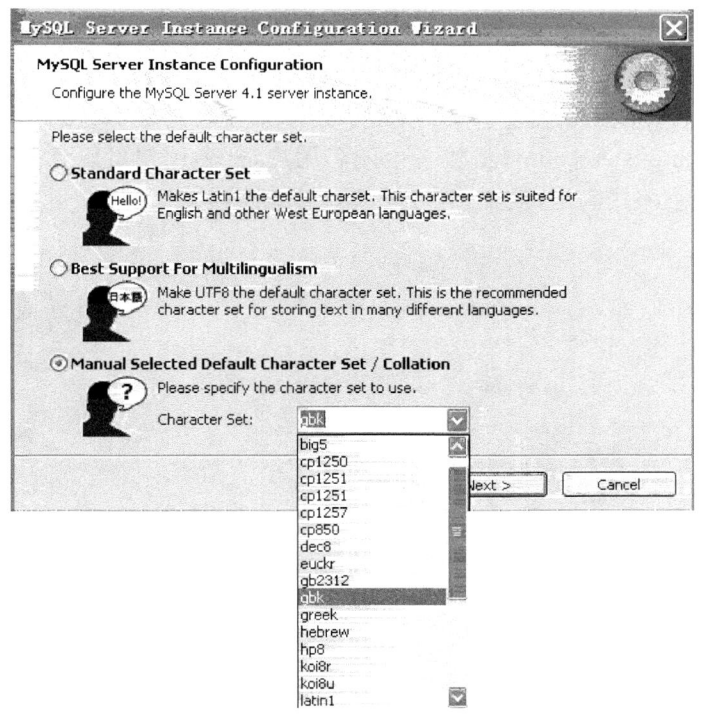

图 5.17　字符集

（18）选择是否将 MySQL 安装为 Windows 服务，还可以指定 Service Name（服务标识名称），是否将 MySQL 的 bin 目录加入 Windows PATH（加入后，就可以直接使用 bin 下的文件，而不用指出目录名,比如连接,"MySQL.exe-uusername-ppassword;"就可以了,不用指出 MySQL.exe 的完整地址,很方便),我这里全部打上了勾,Service Name 不变,按"Next"继续,见图 5.18。

图 5.18　Windows 服务

(19)这一步询问是否要修改默认 root 用户(超级管理)的密码(默认为空),"New root password"如果要修改,就在此填入新密码(如果是重装,并且之前已经设置了密码,在这里更改密码可能会出错,请留空,并将"Modify Security Settings"前面的勾去掉,安装配置完成后另行修改密码),"Confirm(再输一遍)"内再填一次,防止输错。"Enable root access from remote machines (是否允许 root 用户在其他的机器上登录,如果要安全,就不要勾上,如果要方便,就勾上它)"。最后"Create An Anonymous Account(新建一个匿名用户,匿名用户可以连接数据库,不能操作数据,包括查询)",一般就不用勾了,设置完毕,按"Next"继续,见图 5.19。

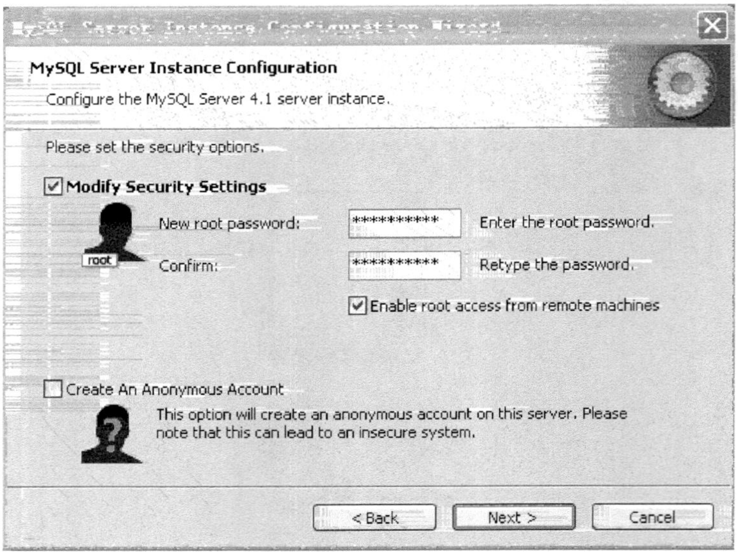

图 5.19　账号密码

(20) 确认设置无误,如果有误,按"Back"返回检查,按"Execute"使设置生效,见图5.20。

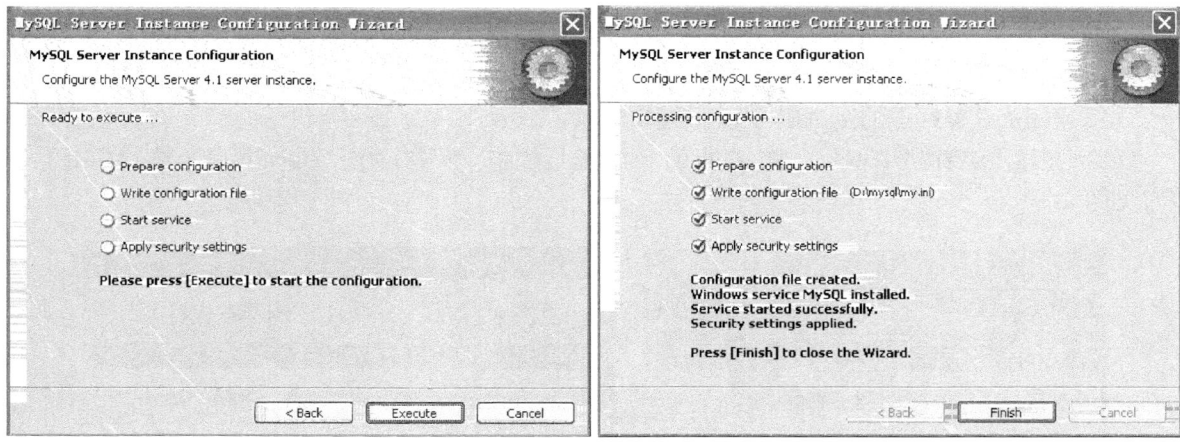

图 5.20 生效　　　　　　　　　　　图 5.21 完成配置

(21) 设置完毕,按"Finish"结束 MySQL 的安装与配置,见图 5.21。

MySQL 安装完成后,为了 PHP 能够访问 MySQL 数据库,打开 php.ini 文件,进行配置,将 extension＝php_MySQL.dll 和 extension＝php_MySQLi.dll 前的分号去掉,见图 5.22。

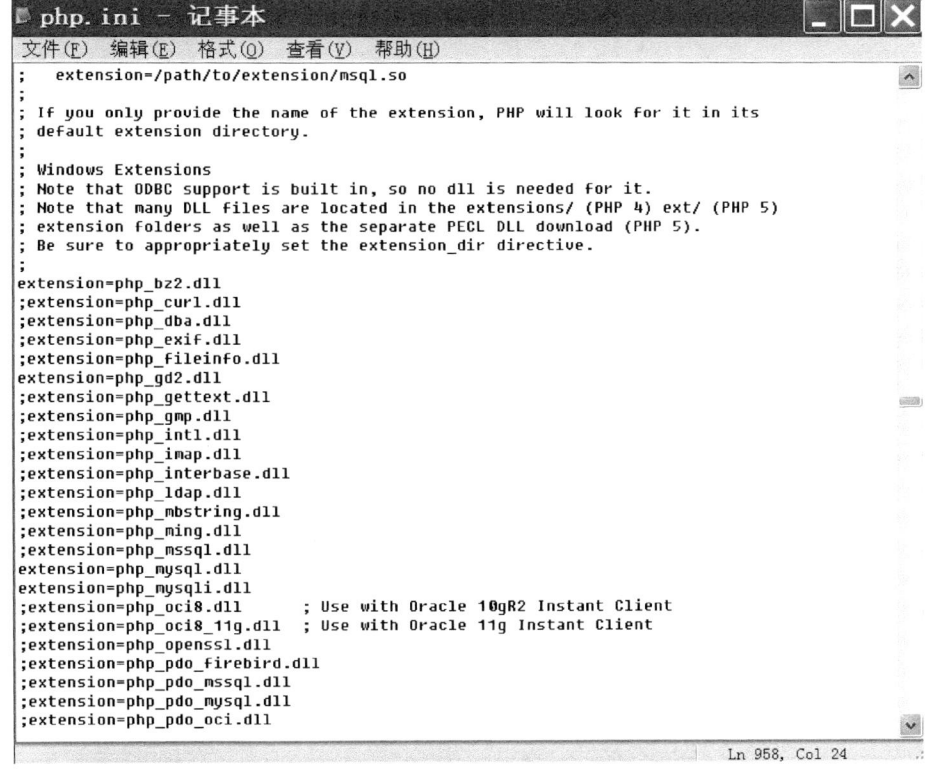

图 5.22 修改 PHP.ini

项目五　MySQL 数据库的安装配置 / 37

任务二　构建 Dreamweaver 下的动态站点

前面的章节，我们完成了 Apache＋php＋MySQL 的安装配置，作为辅助开发工具，需要对 Dreamweaver 进行配置，以达到快速浏览的目的。

（1）点击"设置站点的测试服务器"，见图 5.23。

（2）在站点名称栏中填入名称，站点的 http 地址栏中，填写"http://localhost"，然后点击下一步，见图 5.24。

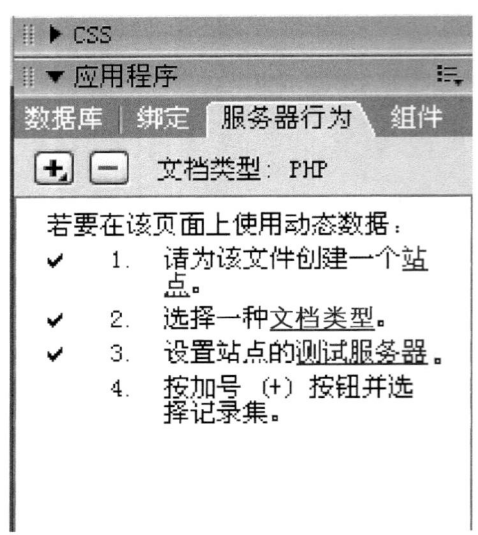

图 5.23　测试服务器

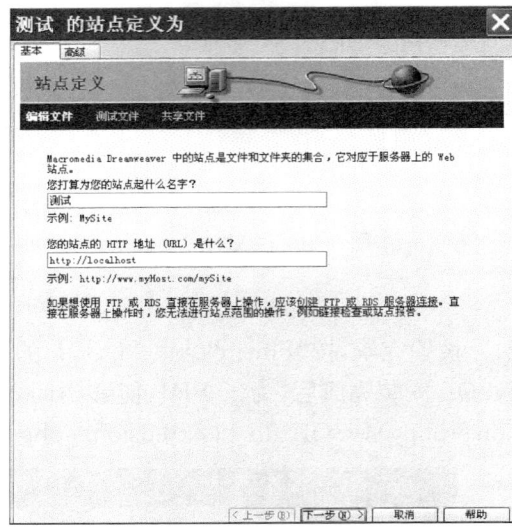

图 5.24　站点名称

（3）选择"是，我想使用服务器技术"，并选择"php MySQL"服务，然后点击下一步，见图 5.25。

（4）选择"在本地进行编辑和测试"，并填写 Apache 发布目录的位置，然后点击下一步，见图 5.26。

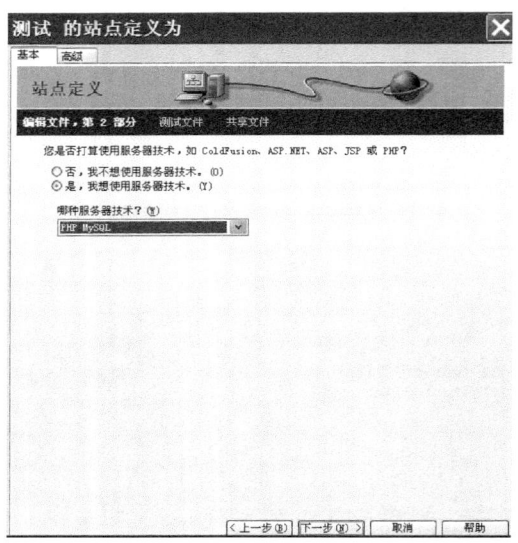

图 5.25　服务器技术选择

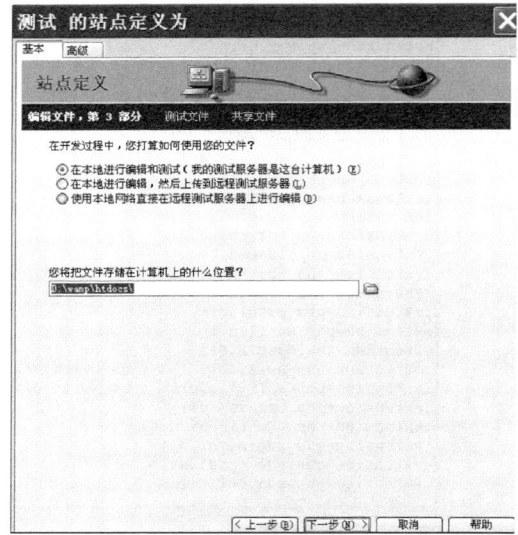

图 5.26　存储位置选择

(5) 填写 url,默认"http://localhsot",见图 5.27。
(6) 远程服务器一项,选择"否",见图 5.28。

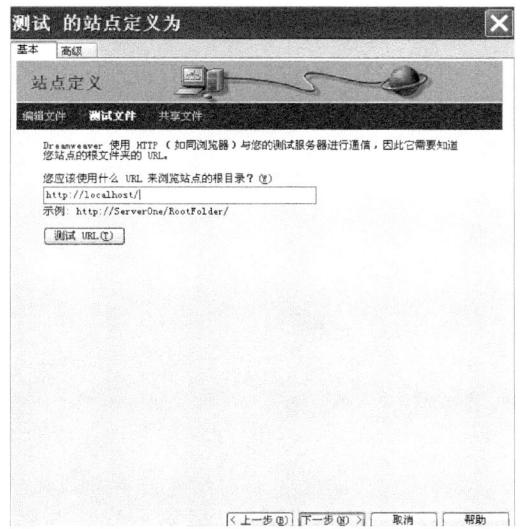

图 5.27 填写 URL   图 5.28 远程服务器选择

(7) PHP 站点配置完成,见图 5.29。

如果配置的过程中出现错误,还可以点击"本地信息"进行修改,见图 5.30。

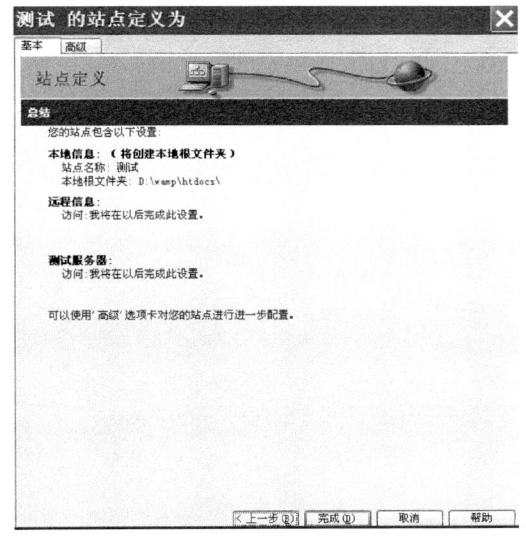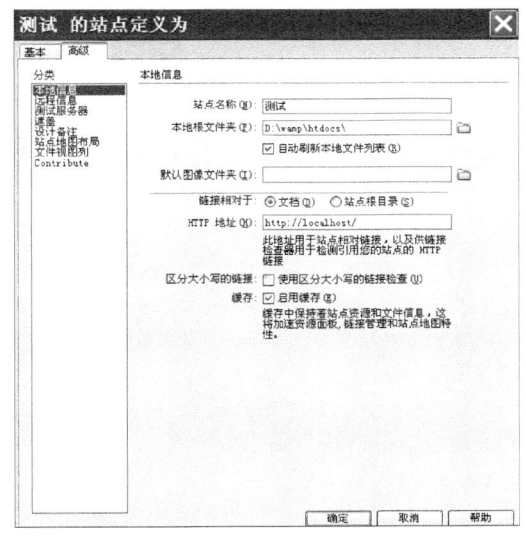

图 5.29 完成配置   图 5.30 修改配置

五、考核标准

(1) 完成 MySQL 的安装。(30 分)
(2) 完成 MySQL 的配置。(20 分)
(3) 正确运行 MySQL、访问数据库。(20 分)
(4) 完成 Dreamweaver 下的服务器配置(30 分)

项目六　PHP 对 MySQL 数据的基本操作

一、实训目的

- 掌握 PHP 连接数据库的方法；
- 掌握 PHP 调用数据库进行数据增加、修改、删除的方法。

二、实训要求

- 使用 PHP，通过网页访问数据库；
- 使用 PHP，向数据库添加数据、修改数据和删除数据。

三、实训设计

PHP 开发环境全部配置完成后，可以开始访问数据库。任何一个网站，都离不开数据库的支持，数据库的基本操作可以分为增加、修改、删除、查询。本实训主要通过 PHP 对数据库的基本操作，锻炼学生的基础开发能力。

四、实训内容

任务一　信息添加

以学生信息添加为例，连接数据库，向数据库中添加新的学生信息，包括学生学号、姓名、性别、出生日期、专业号、系别号、入学年份、学制、政治面貌、密码、照片等信息。

页面 1 代码：

```
<! DOCTYPE html PUBLIC "-//W3C//DTD XHTML 1.0 Transitional//EN"
"http://www.w3.org/TR/xhtml1/DTD/xhtml1-transitional.dtd">
<html xmlns="http://www.w3.org/1999/xhtml">
<head>
<meta http-equiv="Content-Type" content="text/html; charset=gb2312" />
<title>录入成绩</title>
</head>

<body>
<form action="add_student.php" target="d2r2c1" method="POST" enctype="multipart/form-data" name="mainForm" id="mainForm">
    <div align="center">增加学生信息</div><br>
    <table width="300" border="1" align="center" bgcolor="#CCFFFF" cellpadding="0" cellspacing="0">
        <tr><td width="85">学号：</td>
            <td width="199">
<input name="学号" type="text" size="20" maxlength="20"></td>
```

```html
        </tr>
        <tr><td>姓名:</td>
          <td>
<input name="姓名" type="text" size="20" maxlength="20"></td>
        </tr>
        <tr><td>性别:</td>
          <td>
<input name="性别" type="text" size="20" maxlength="20"></td>
        </tr>
        <tr><td>出生日期:</td>
          <td>
<input name="出生日期" type="text" size="20" maxlength="20"></td>
        </tr>
        <tr><td>班级号:</td>
          <td>
<input name="班级号" type="text" size="20" maxlength="20"></td>
        </tr>
        <tr><td>专业号:</td>
          <td>
<input name="专业号" type="text" size="20" maxlength="20"></td>
        </tr>
        <tr><td>系别号:</td>
          <td>
<input name="系别号" type="text" size="20" maxlength="20"></td>
        </tr>
        <tr><td>入学年份:</td>
          <td>
<input name="入学年份" type="text" size="20" maxlength="20"></td>
        </tr>
        <tr><td>学制:</td>
          <td>
<input name="学制" type="text" size="20" maxlength="20"></td>
        </tr>
        <tr><td>政治面貌:</td>
          <td>
<input name="政治面貌" type="text" size="20" maxlength="20"></td>
        </tr>
        <tr><td>联系电话:</td>
          <td>
<input name="联系电话" type="text" size="20" maxlength="20"></td>
        </tr>
        <tr><td>密码:</td>
          <td>
<input name="密码" type="text" size="20" maxlength="20"></td>
        </tr>
        <tr><td>照片:</td>
          <td>
<img src="" name="myphoto" /><br>
<input type="file" name="myFile" onchange="mainForm.myphoto.src=this.value;" /><br />
          </td></tr>
```

```html
    </table>
    <br>
    <div align="center">
    <input name="submit" type="submit" value="确定">
        <input name="reset" type="reset" value="取消">
        </div>
        </form>
    </body>
    </html>
```

页面 2 代码：

```php
<?php
    $学号 = $_POST["学号"];
    $姓名 = $_POST["姓名"];
    $性别 = $_POST["性别"];
    $出生日期 = $_POST["出生日期"];
    $班级号 = $_POST["班级号"];
    $专业号 = $_POST["专业号"];
    $系别号 = $_POST["系别号"];
    $入学年份 = $_POST["入学年份"];
    $学制 = $_POST["学制"];
    $政治面貌 = $_POST["政治面貌"];
    $联系电话 = $_POST["联系电话"];
    $密码 = $_POST["密码"];
    $photoname = $_FILES['myFile']['tmp_name'];//上传至服务器后读取
if (!empty($photoname)) {
    $photo = fread(fopen($photoname,"r"),filesize($photoname));//读取图片
    $photo = '0x'. bin2hex($photo);}
$link = mysql_connect("127.0.0.1","root","123456")
    or die("数据库服务器连接失败！<BR>");
mysql_select_db("education", $link) or die("数据库选择失败！<BR>");
mysql_query("set names 'gbk'");
$sql = "select 学号 from 学生表 where 学号 = '$学号'";
$result = mysql_query($sql, $link);
$row = mysql_fetch_array($result);
$sql = "INSERT INTO '学生表' VALUES ('$学号','$姓名','$性别','$出生日期','$班级号','$专业号','$系别号','$入学年份','$学制','$政治面貌','$联系电话','$密码',$photo);";
if (mysql_query($sql, $link))
    echo "信息增加成功！";
else
    echo '信息增加失败！';
?>
```

代码运行效果见图 6.1。

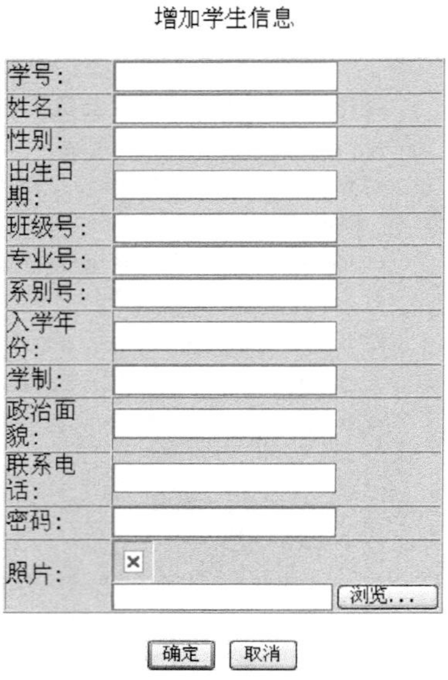

图 6.1 添加学生

任务二 信息修改

在系统的使用过程中,经常需要修改用户信息,本实训以教师信息为例,进行修改,其中教师编号为主键。

页面1代码:

```
<!DOCTYPE html PUBLIC "-//W3C//DTD XHTML 1.0 Transitional//EN"
"http://www.w3.org/TR/xhtml1/DTD/xhtml1-transitional.dtd">
<html xmlns="http://www.w3.org/1999/xhtml">
<head>
<meta http-equiv="Content-Type" content="text/html; charset=gb2312" />
<title>修改信息</title>
</head>

<body>
<form action="update_teacher_edit.php" target="d2r2c1" method="get">
<b>请输入欲修改的教师编号:</b>
<input type="text" name="教师编号">
<input type="submit" value="查询">
<input type="reset" value="取消">
</form>
</body>
</html>
```

页面2代码:

```
<html>
<head>
<meta http-equiv="Content-Type" content="text/html; charset=gb2312">
<title>班级编辑</title>
</head>
<body>
<?php
  $教师编号 = $_GET["教师编号"];
  if ($教师编号 == "") {
    echo "教师编号代码不能为空!";
    die();
  }
  $link = mysql_connect("127.0.0.1","root","123456")
    or die("数据库服务器连接失败! <BR>");
  mysql_select_db("education", $link)
    or die("数据库选择失败! <BR>");
  mysql_query("set names 'gbk'");
  $sql = "select 教师编号,教师姓名,性别,系别号,学历,联系电话,密码 from 教师表 where 教师编号='$教师编号'";
  $result = mysql_query($sql, $link);
  $row = mysql_fetch_array($result);
  if (!$row) {
    echo "无此教师编号!";
    die();
```

```php
          }
        $教师编号 = $row["教师编号"];
        $教师姓名 = $row["教师姓名"];
        $性别 = $row["性别"];
        $系别号 = $row["系别号"];
        $学历 = $row["学历"];
        $联系电话 = $row["联系电话"];
        $密码 = $row["密码"];
    ?>
    <form action="update_teacher.php" method="get">
      <div align="center">修改教师信息</div>   <br>
      <table width="300" border="3" cellpadding="0" cellspacing="0" bordercolor="#99FFFF" align="center"  bgcolor="#CDE0F1">
        <tr><td width="85">教师编号：</td>
          <td width="199"><input name="教师编号" type="text" value="<?php echo $教师编号;?>" size="9" maxlength="9"></td>
        </tr>
        <tr>   <td>教师姓名：</td>
          <td><input name="教师姓名" type="text" value="<?php echo $教师姓名;?>" size="15" maxlength="15"></td>
        </tr>
        <tr>   <td>性别：</td>
          <td><input name="性别" type="text" value="<?php echo $性别;?>" size="20" maxlength="30"></td>
        </tr>
        <tr>   <td>系别号：</td>
          <td><input name="系别号" type="text" value="<?php echo $系别号;?>" size="20" maxlength="30"></td>
        </tr>
        <tr>   <td>学历：</td>
          <td><input name="学历" type="text" value="<?php echo $学历;?>" size="20" maxlength="30"></td>
        </tr>
        <tr>   <td>联系电话：</td>
          <td><input name="联系电话" type="text" value="<?php echo $联系电话;?>" size="20" maxlength="30"></td>
        </tr>
        <tr>   <td>密码：</td>
          <td><input name="密码" type="text" value="<?php echo $密码;?>" size="20" maxlength="30"></td>
        </tr>
      </table>
      <input name="教师编号0"  type="hidden" value="<?php echo $教师编号;?>">
      <br>
      <div align="center">
        <input name="submit" type="submit" value="确定">
        <input name="reset" type="reset" value="取消">
      </div>
    </form>
  </body>
</html>
```

页面 3 代码：

```php
<? php
  $教师编号=$_GET["教师编号"];
  $教师姓名=$_GET["教师姓名"];
  $性别=$_GET["性别"];
  $系别号=$_GET["系别号"];
  $学历=$_GET["学历"];
  $联系电话=$_GET["联系电话"];
  $密码=$_GET["密码"];
  $教师编号0=$_GET["教师编号0"];

$link=mysql_connect("127.0.0.1","root","123456")
    or die("数据库服务器连接失败！<BR>");
  mysql_select_db("education",$link) or die("数据库选择失败！<BR>");
  mysql_query("set names 'gbk'");
  if($教师编号!=$教师编号0){
    $sql="select 教师编号 from 教师表 where 教师编号='$教师编号'";
    $result=mysql_query($sql,$link);
    $row = mysql_fetch_array($result);
    if($row){
       echo "此教师编号已经存在！";
       die();
    }
  }
  $sql="update 教师表 set 教师编号='$教师编号',教师姓名='$教师姓名'";
  $sql=$sql.",性别='$性别',系别号='$系别号',学历='$学历',联系电话='$联系电话',密码='$密码' where 教师编号='$教师编号0'";
  if(mysql_query($sql,$link))
    echo "教师信息修改成功！";
  else
    echo '教师信息修改失败！';
?>
```

代码运行效果见图 6.2。

图 6.2　修改信息

任务三　信 息 删 除

系统中的无用数据，需要进行清理，本实训以班级表删除为例，讲解如何使用 PHP 删除数据库中的数据。

页面 1 代码：

```
<html>
<head>
<meta http-equiv="Content-Type" content="text/html; charset=gb2312">
<title>班级删除</title>
</head>
```

```html
<body>
<form action="class_delete.php" method="get">
    请输入欲删除班级的代码：
    <input name="bjdm" type="text" size="9" maxlength="9">
    <input name="submit" type="submit" value="确定">
    <input name="reset" type="reset" value="取消">
</form>
</body>
</html>
```

页面2代码：

```php
<?php
  $bjdm=trim($bjdm);
  if($bjdm==""){
      echo "班级代码不能为空!";
      die();
  }
  $link=mysql_connect("localhost","root","123456")
      or die("数据库服务器连接失败!<BR>");
  mysql_select_db("student",$link) or die("数据库选择失败!<BR>");
  mysql_query("set names 'gbk'");
  $sql="select bjdm from t_class where bjdm='$bjdm'";
  $result=mysql_query($sql,$link);
  $row = mysql_fetch_array($result);
  if(!$row){
      echo "无此班级代码!";
      die();
  }
  $sql="delete from t_class where bjdm='$bjdm'";
  if(mysql_query($sql,$link))
      echo "班级删除成功!";
  else
      echo '班级删除失败!';
?>
```

代码运行效果见图6.3。

图 6.3　删除信息

五、考核标准

(1) 使用PHP正确访问数据库服务器。(10分)

(2) 完成数据的增加。(30分)

(3) 完成数据的修改。(30分)

(4) 完成数据的删除。(30分)

项目七 PHP 对 MySQL 数据的查询

一、实训目的

- 掌握 PHP 查询数据库的基本方法；
- 掌握 PHP 分页显示功能；
- 掌握 PHP 多条件查询功能。

二、实训要求

- 使用 PHP，通过网页访问并查询数据库中的数据；
- 使用 PHP，进行数据多条件查询，并分页显示。

三、实训设计

网站开发过程中，需要从数据库查询数据，生成网页。因此 PHP 对数据库的查询操作是最常用的操作之一。本实训通过不同的查询方式和不同的查询内容，锻炼学生的数据查询能力以及页面自动生成能力。

四、实训内容

任务一 数据基本查询

使用 PHP 访问数据库，并将数据显示在一个 HTML 表格中了。

```php
<? php
$con = mysql_connect("localhost","root","");
if (! $con)
  {
  die('数据库连接失败：'. mysql_error());
  }
  else
  {
  mysql_query("SET NAMES UTF8");
  mysql_query("set character_set_client=utf8");
  mysql_query("set character_set_results=utf8");
  mysql_select_db("demosql", $con);
  $result = mysql_query("SELECT * FROM teacher");
  //在表格中输出显示结果
  echo "<table border='1'>
<tr>
<th>id</th>
<th>name</th>
</tr>";
```

```php
    while( $row = mysql_fetch_array( $result))
    {
echo "<tr>";
echo "<td>" . $row['id'] . "</td>";
echo "<td>" . $row['name'] . "</td>";
echo "</tr>";
    }
    echo "</table>";
}
mysql_close( $con);
?>
```

任务二 数据分页显示

分页是目前在显示大量结果时所采用的最好的方式。几乎在每一个 Web 应用程序都需要划分返回的数据,并按页显示。要求使用 PHP 访问数据库,并将结果集以分页形式显示。

```php
<?php
/*
    Place code to connect to your DB here.
*/
include('config.php');      // include your code to connect to DB.
$tbl_name = "";             //your table name
// How many adjacent pages should be shown on each side?
$adjacents = 3;

/*
    First get total number of rows in data table.
    If you have a WHERE clause in your query, make sure you mirror it here.
*/
$query = "SELECT COUNT(*) as num FROM $tbl_name";
$total_pages = mysql_fetch_array(mysql_query( $query));
$total_pages = $total_pages[num];
/* Setup vars for query. */
$targetpage = "filename.php";   //your file name (the name of this file)
$limit = 2;
$page = $_GET['page'];
if( $page)
    $start = ( $page - 1) * $limit;           //first item to display on this page
else
    $start = 0;                               //if no page var is given, set start to 0
/* Get data. */
$sql = "SELECT column_name FROM $tbl_name LIMIT $start, $limit";
$result = mysql_query( $sql);
/* Setup page vars for display. */
if ( $page == 0) $page = 1;                   //If no page var is given, default to 1.
$prev = $page - 1;
$lastpage = ceil( $total_pages/ $limit);      //lastpage is = total pages / items per page, rounded up.
```

```php
$lpm1 = $lastpage - 1;
//last page minus 1
/*
    Now we apply our rules and draw the pagination object.
    We're actually saving the code to a variable in case we want to draw it more than once.
*/
$pagination = "";
if($lastpage > 1)
{
    $pagination .= "<div class=\"pagination\">";
    //previous button
    if ($page > 1)
        $pagination.= "<a href=\"$targetpage?page=$prev\">◆ previous</a>";
    else
        $pagination.= "<span class=\"disabled\">◆ previous</span>";

    //pages
    if ($lastpage < 7 + ($adjacents * 2))    //not enough pages to bother breaking it up
    {
        for ($counter = 1; $counter <= $lastpage; $counter++)
        {
            if ($counter == $page)
                $pagination.= "<span class=\"current\">$counter</span>";
            else
                $pagination.= "<a href=\"$targetpage?page=$counter\">$counter</a>";
        }
    }
    elseif($lastpage > 5 + ($adjacents * 2))    //enough pages to hide some
    {
        //close to beginning; only hide later pages
        if($page < 1 + ($adjacents * 2))
        {
            for ($counter = 1; $counter < 4 + ($adjacents * 2); $counter++)
            {
                if ($counter == $page)
                    $pagination.= "<span class=\"current\">$counter</span>";
                else
                    $pagination.= "<a href=\"$targetpage?page=$counter\">$counter</a>";
            }
            $pagination.= "...";
            $pagination.= "<a href=\"$targetpage?page=$lpm1\">$lpm1</a>";
            $pagination.= "<a href=\"$targetpage?page=$lastpage\">$lastpage</a>";
        }
        //in middle; hide some front and some back
        elseif($lastpage - ($adjacents * 2) > $page && $page > ($adjacents * 2))
        {
            $pagination.= "<a href=\"$targetpage?page=1\">1</a>";
```

```php
                    $pagination .= "<a href=\"$targetpage?page=2\">2</a>";
                    $pagination .= "...";
                    for ($counter = $page - $adjacents; $counter <= $page + $adjacents; $counter++)
                    {
                        if ($counter == $page)
                            $pagination .= "<span class=\"current\">$counter</span>";
                        else
                            $pagination .= "<a href=\"$targetpage?page=$counter\">$counter</a>";
                    }
                    $pagination .= "...";
                    $pagination .= "<a href=\"$targetpage?page=$lpm1\">$lpm1</a>";
                    $pagination .= "<a href=\"$targetpage?page=$lastpage\">$lastpage</a>";
                }
                //close to end; only hide early pages
                else
                {
                    $pagination .= "<a href=\"$targetpage?page=1\">1</a>";
                    $pagination .= "<a href=\"$targetpage?page=2\">2</a>";
                    $pagination .= "...";
                    for ($counter = $lastpage - (2 + ($adjacents * 2)); $counter <= $lastpage; $counter++)
                    {
                        if ($counter == $page)
                            $pagination .= "<span class=\"current\">$counter</span>";
                        else
                            $pagination .= "<a href=\"$targetpage?page=$counter\">$counter</a>";
                    }
                }
            }

            //next button
            if ($page < $counter - 1)
                $pagination .= "<a href=\"$targetpage?page=$next\">next ❥</a>";
            else
                $pagination .= "<span class=\"disabled\">next ❥</span>";
            $pagination .= "</div>\n";
        }
?>

        <?php
        while($row = mysql_fetch_array($result))
        {

        // Your while loop here

        }
```

```
    ?>
<?=$pagination?>
```

代码运行效果见图7.1。

图 7.1 分页导航

任务三 创建分页类

分页代码一般较长,并且在系统的很多页面中都会用到,为了降低代码书写量,提高代码复用,可以创建分页类。要求使用 PHP 创建分页类。

分页类代码:

```
<?php
/*
 * PHP 分页类
 * @package Page
 * @Created 2013-03-27
 * @Modify  2013-03-27
 * @link http://www.60ie.net
 * Example:
        $myPage=new Pager(1300,intval($CurrentPage));
        $pageStr= $myPage->GetPagerContent();
        echo $pageStr;
*/
class Pager{
    private $pageSize = 10;
    private $pageIndex;
    private $totalNum;
    private $totalPagesCount;
    private $pageUrl;
    private static $_instance;
    public function __construct($p_totalNum, $p_pageIndex, $p_pageSize = 10, $p_initNum=3, $p_initMaxNum=5) {
        if (!isset ($p_totalNum) || !isset($p_pageIndex)) {
            die ( "pager initial error" );
        }
        $this->totalNum = $p_totalNum;
        $this->pageIndex = $p_pageIndex;
        $this->pageSize = $p_pageSize;
        $this->initNum= $p_initNum;
        $this->initMaxNum= $p_initMaxNum;
        $this->totalPagesCount= ceil($p_totalNum / $p_pageSize);
        $this->pageUrl= $this->_getPageUrl();
        $this->_initPagerLegal();
```

```php
        }
    /**
     * 获取去除page部分的当前URL字符串
     *
     * @return String URL字符串
     */
    private function _getPageUrl() {
        $CurrentUrl = $_SERVER["REQUEST_URI"];
        $arrUrl     = parse_url($CurrentUrl);
        $urlQuery   = $arrUrl["query"];
        if($urlQuery){
            $urlQuery = ereg_replace("(^|&)page=" . $this->pageIndex, "", $urlQuery);
            $CurrentUrl = str_replace($arrUrl["query"], $urlQuery, $CurrentUrl);
            if($urlQuery){
                $CurrentUrl .= "&page";
            }
            else $CurrentUrl .= "page";
        } else {
            $CurrentUrl .= "?page";
        }
        return $CurrentUrl;
    }
    /*
     * 设置页面参数合法性
     * @return void
     */
    private function _initPagerLegal()
    {
        if((!is_numeric($this->pageIndex)) || $this->pageIndex<1)
        {
            $this->pageIndex = 1;
        }elseif($this->pageIndex > $this->totalPagesCount)
        {
            $this->pageIndex = $this->totalPagesCount;
        }
    }
    //{$this->pageUrl}={$i}
    //{$this->CurrentUrl}={$this->TotalPages}
    public function GetPagerContent() {
        $str = "<div class=\"Pagination\">";
        //首页 上一页
        if($this->pageIndex == 1)
        {
            $str .= "<a href='JavaScript:void(0)' class='tips' title='首页'>首页</a> "."\n";
            $str .= "<a href='javascript:void(0)' class='tips' title='上一页'>上一页</a> "."\n"."\n";
        }else
        {
            $str .= "<a href='{$this->pageUrl}=1' class='tips' title='首页'>首页</a>
```

```php
        "."\n";
                            $str .= "<a href='{$this->pageUrl}=".($this->pageIndex-1)."' class='tips' title='上一页'>上一页</a>"."\n"."\n";
        }
        /*
        除首末后 页面分页逻辑
        */
        //10页(含)以下
        $currnt="";
        if($this->totalPagesCount<=10)
        {
            for($i=1;$i<=$this->totalPagesCount;$i++)
            {
                if($i==$this->pageIndex)
                {   $currnt=" class='current'";}
                else
                {   $currnt="";    }
                $str .= "<a href='{$this->pageUrl}={$i}'{$currnt}>$i</a>"."\n";
            }
        }else                                       //10页以上
        {   if($this->pageIndex<3)   //当前页小于3
            {
                for($i=1;$i<=3;$i++)
                {
                    if($i==$this->pageIndex)
                    {   $currnt=" class='current'";}
                    else
                    {   $currnt="";   }
                    $str .= "<a href='{$this->pageUrl}={$i}'{$currnt}>$i</a>"."\n";
                }
                $str.= "<span class=\"dot\">……</span>"."\n";
                for($i=$this->totalPagesCount-3+1;$i<=$this->totalPagesCount;$i++)//功能1
                {
                    $str .= "<a href='{$this->pageUrl}={$i}'>$i</a>"."\n";
                }
            }elseif($this->pageIndex<=5)    // 5>=当前页>=3
            {
                for($i=1;$i<=($this->pageIndex+1);$i++)
                {
                    if($i==$this->pageIndex)
                    {   $currnt=" class='current'";}
                    else
                    {   $currnt="";   }
                    $str .= "<a href='{$this->pageUrl}={$i}'{$currnt}>$i</a>"."\n";
                }
                $str .= "<span class=\"dot\">……</span>"."\n";
```

```php
                    for($i=$this->totalPagesCount-3+1;$i<=$this->totalPagesCount;$i++)//功能1
                    {
                        $str .="<a href='{$this->pageUrl}={$i}'>$i</a>"."\n";
                    }
                }elseif(5<$this->pageIndex  && $this->pageIndex<=$this->totalPagesCount-5)         //当前页大于5,同时小于总页数-5
                {
                    for($i=1;$i<=3;$i++)
                    {
                        $str .="<a href='{$this->pageUrl}={$i}'>$i</a>"."\n";
                    }
                    $str.="<span class=\"dot\">……</span>";
                    for($i=$this->pageIndex-1;$i<=$this->pageIndex+1 && $i<=$this->totalPagesCount-5+1;$i++)
                    {
                        if($i==$this->pageIndex)
                        {   $currnt=" class='current'";}
                        else
                        {   $currnt="";   }
                        $str .="<a href='{$this->pageUrl}={$i}' {$currnt}>$i</a>"."\n";
                    }
                    $str.="<span class=\"dot\">……</span>";
                    for($i=$this->totalPagesCount-3+1;$i<=$this->totalPagesCount;$i++)
                    {
                        $str .="<a href='{$this->pageUrl}={$i}'>$i</a>"."\n";
                    }
                }else
                {
                    for($i=1;$i<=3;$i++)
                    {
                        $str .="<a href='{$this->pageUrl}={$i}'>$i</a>"."\n";
                    }
                    $str.="<span class=\"dot\">……</span>"."\n";
                    for($i=$this->totalPagesCount-5;$i<=$this->totalPagesCount;$i++)//功能1
                    {
                        if($i==$this->pageIndex)
                        {   $currnt=" class='current'";}
                        else
                        {   $currnt="";   }
                        $str .="<a href='{$this->pageUrl}={$i}' {$currnt}>$i</a>"."\n";
                    }
                }
            }
            /*
                除首末后 页面分页逻辑结束
            */
```

```php
            //下一页 末页
            if($this->pageIndex == $this->totalPagesCount)
            {
                $str .= "\n"."<a href='javascript:void(0)' class='tips' title='下一页'>下一页</a>"."\n";
                $str .= "<a href='javascript:void(0)' class='tips' title='末页'>末页</a>"."\n";
            }else
            {
                $str .= "\n"."<a href='{$this->pageUrl}=".($this->pageIndex+1)."' class='tips' title='下一页'>下一页</a>"."\n";
                $str .= "<a href='{$this->pageUrl}={$this->totalPagesCount}' class='tips' title='末页'>末页</a>"."\n";
            }

            $str .= "</div>";
            return $str;
        }
    /**
     * 获得实例
     * @return
     */
//    static public function getInstance() {
//        if (is_null ( self::$_instance )) {
//            self::$_instance = new pager ();
//        }
//        return self::$_instance;
//    }
}
?>  代码可以让你的开发很有帮助。
```

调用的分页类代码：

```
<head>
<meta http-equiv="Content-Type" content="text/html; charset=utf-8" />
<title>----分页演示-----</title>
<link href="pager.css" type="text/css" rel="stylesheet" />
</head>
<body>
    <?php
    include "pager.class.php";
    $CurrentPage = isset($_GET['page'])? $_GET['page']:1;
    //die($CurrentPage);
    $myPage = new pager(1300,intval($CurrentPage));
      $pageStr= $myPage->GetPagerContent();
    //echo $pageStr;
    $myPage = new pager(90,intval($CurrentPage));
    $pageStr= $myPage->GetPagerContent();
    echo $pageStr;
    ?>
```

```
</body>
</html>
```

分页效果美化代码：

```css
body,html{ padding:0px; margin:0px; color:#333333; font-family:"宋体",Arial,Lucida,Verdana,Helvetica,sans-serif; font-size:12px; line-height:150%;}

h1,h2,h3,h4,h5,h6,ul,li,dl,dt,dd,form,img,p,label{margin:0; padding:0; border:none; list-style-type:none;}
/**前台分页样式**/
.Pagination {margin:10px 0 0; padding:5px 0; text-align:rightright; height:20px; line-height:20px; font-family:Arial, Helvetica, sans-serif,"宋体";}
.Pagination a {margin-left:2px;padding:2px 7px 2px;}
.Pagination .dot{ border:medium none; padding:4px 8px}
.Pagination a:link, .Pagination a:visited {border:1px solid #dedede;color:#696969;text-decoration:none;}
.Pagination a:hover, .Pagination a:active, .Pagination a.current:link, .Pagination a.current:visited {border:1px solid #dedede;color:#fff; background-color:#ff6600; background-image:none; border:#ff6600 solid 1px;}
.Pagination .selectBar{ border:#dedede solid 1px; font-size:12px; width:95px; height:21px; line-height:21px; margin-left:10px; display:inline}
.Pagination a.tips{_padding:4px 7px 1px;}
```

任务四　多条件数据查询

在很多情况下，我们需要输入很多条件以精确查询数据，比如，要查询某个班级的某个学生。

页面1代码：

```html
<html>
<head>
<title>教师基本信息</title>
</head>
<body>
<form action="teacher_select_score.php" target="d2r2c1" method="get">
<b>请输入班级号：</b>
<input type="text" name="班级号">
<b>请输入课程号：</b>
<input type="text" name="课程号">

<input type="submit" value="查询">
<input type="reset" value="取消">
</form>
</body>
</html>
```

页面2代码：

```html
<html>
<head>
```

```php
<meta http-equiv="Content-Type" content="text/html; charset=gb2312">
<title>学生查询结果</title>
</head>
<body>
<?php
    $班级号=$_GET["班级号"];
    $课程号=$_GET["课程号"];

if($班级号==""||$课程号==""){
echo "班级号,课程号不能为空!";
die();
}
    $link=mysql_connect("127.0.0.1","root","123456") or die("数据库服务器连接失败!<br>");

mysql_select_db("education",$link) or die("数据库选择成功!<br>");
    mysql_query("set names 'gbk'");
    $sql="select 成绩表.学号,学生表.姓名,学生表.班级号,成绩表.课程号,课程信息表.课程名,成绩 from 学生表,课程信息表,成绩表 where 学生表.学号=成绩表.学号 and  课程信息表.课程号=成绩表.课程号 and   学生表.班级号='$班级号' and 成绩表.课程号='$课程号'" ;
    $result=mysql_query($sql,$link) or die("有错<br>");
  $rows=mysql_num_rows($result);
    if ($rows==0)   {
      echo "没有满足条件的记录!";
      die();
    }
    $pagesize=10;   //每页的记录数(在此暂设为5,通常应设为10)
    $pagecount=ceil($rows/$pagesize);  //总页数
    //$pageno 的值为当前页的页号
    if (!isset($pageno)||$pageno<1)
       $pageno=1;
    if ($pageno>$pagecount)
       $pageno=$pagecount;
       $offset=($pageno-1)*$pagesize;
mysql_data_seek($result,$offset);
?>
<div align="center"><strong>成绩查询结果</strong> </div>
<table style="table-layout:fixed;"  width="90%" border="3" cellpadding="0" cellspacing="0" bordercolor="#99FFFF" align="center"  bgcolor="#CDE0F1">
<thead bgcolor="#3399FF">
  <tr>
    <td><div align="center">学号</div></td>
    <td><div align="center">姓名</div></td>
    <td><div align="center">班级号</div></td>
    <td><div align="center">课程号</div></td>
    <td><div align="center">课程名</div></td>
    <td><div align="center">成绩</div></td>
    <td><div align="center">操作</div></td>
  </tr>
  </thead>
<?php
    $i=0;
```

```php
      while($row=mysql_fetch_array($result))  {
?>
  <tr>
    <td><div align="center"><?php echo $row['学号']; ?></div></td>
    <td><div align="center"><?php echo $row['姓名']; ?></div></td>
    <td><div align="center"><?php echo $row['班级号']; ?></div></td>
    <td><div align="center"><?php echo $row['课程号']; ?></div></td>
    <td><div align="center"><?php echo $row['课程名']; ?></div></td>
    <td><div align="center"><?php echo $row['成绩']; ?></div></td>
    <td><div align="center">
      <a href="teacher_select_detail2.php?学号=<?php echo $row['学号']; ?>" target="_self">详情</a>
    </div></td>
  </tr>

<?php
    $i=$i+1;
    if ($i==$pagesize)
      break;
    }


?>
</table>
<div align="center">
[第<?php echo $pageno; ?>页/共<?php echo $pagecount; ?>页]
<?php
$href=$PHP_SELF."?班级号=".urlencode($班级号);
$href=$PHP_SELF."?课程号=".urlencode($课程号);
if ($pageno<>1) {
?>
  <a href="<?php echo $href; ?>&pageno=1">首页</a>
  <a href="<?php echo $href; ?>&pageno=<?php echo $pageno-1; ?>">上一页</a>
<?php
}
if ($pageno<> $pagecount) {
?>
<a href="<?php echo $href; ?>&pageno=<?php echo $pageno+1; ?>">下一页</a>
<a href="<?php echo $href; ?>&pageno=<?php echo $pagecount; ?>">尾页</a>
<?php
}
?>
[共找到<?php echo $rows; ?>个记录]
</div>
</body>
</html>
```

代码运行效果见图 7.2。

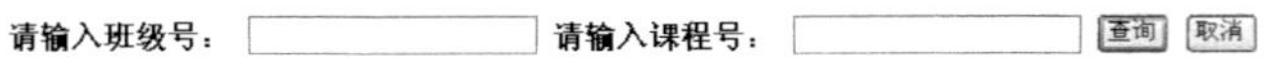

图 7.2 多条件查询

五、考核标准

（1）版面布局合理清晰，整体效果美观，观赏性强。（20分）
（2）网页中没有明显的错误（如超链接、图片无法显示、错别字等）。（10分）
（3）完成基本数据查询。（20分）
（4）完成分页显示。（30分）
（5）完成多条件查询。（20分）

项目八 用户登录控制

一、实训目的

- 掌握 PHP session 的基本用法；
- 掌握用户账号密码的验证；
- 掌握多重身份多重权限的登录控制。

二、实训要求

- 完成 PHP session 在页面中的启用、查询、关闭；
- 完成连接数据库的用户登录验证；
- 完成连接数据库的多重身份验证。

三、实训设计

登录功能是网站日常使用率很高的功能之一，任何用户在不同网站都会有不同的账号。而同一个网站也会有不同级别的账号，比如管理员账号、普通用户账号、VIP 账号等。本实训通过 PHP session 配置、基本登录功能、多用户登录功能等方面训练学生，提高学生关于登录控制方面的能力。

四、实训内容

任务一 PHP session 的配置

php.ini 中有关 session 的一些设定会影响到 session 函数的使用，现在以 php5 版本为例，我们来了解一下 php.ini 中有关 session 的设定：

➢ session.save_handler = "files"

存储和检索与会话关联的数据的处理器名字，默认为文件("files")。如果想要使用自定义的处理器(如基于数据库的处理器)，可用"user"。

➢ session.save_path = "/tmp"

传递给存储处理器的参数。对于 files 处理器，此值是创建会话数据文件的路径。Windows 下默认为临时文件夹路径。你可以使用"N;[MODE;]/path"这样模式定义该路径(N 是一个整数)。N 表示使用 N 层深度的子目录，而不是将所有数据文件都保存在一个目录下。[MODE;]可选，必须使用 8 进制数，默认 600(=384)，表示每个目录下最多保存的会话文件数量。这是一个提高大量会话性能的好主意。注意 0；"N;[MODE;]/path"两边的双引号不能省略。注意：如果该文件夹可以被不安全的用户访问(比如默认的"/tmp")，那么将会带来安全漏洞。

➢ session.name = "PHPSESSID"

用在 cookie 里的会话 ID 标识名，只能包含字母和数字。

➢ session.auto_start = Off

在客户访问任何页面时都自动初始化会话,默认禁止。因为类定义必须在会话启动之前被载入,所以若打开这个选项,你就不能在会话中存放对象。

- session.serialize_handler = "php"

用来序列化/解序列化数据的处理器,PHP 是标准序列化/解序列化处理器。另外还可以使用 "php_binary"。当启用了 WDDX 支持以后,将只能使用 "wddx"。

- session.gc_maxlifetime = 1440

设定保存的 session 文件生存期,超过此参数设定秒数后,保存的数据将被视为垃圾并由垃圾回收程序清理。判断标准是最后访问数据的时间(对于 FAT 文件系统是最后刷新数据的时间)。如果多个脚本共享同一个 session.save_path 目录但 session.gc_maxlifetime 不同,将以所有 session.gc_maxlifetime 指令中的最小值为准。

如果你在 session.save_path 选项中设定使用子目录来存储 session 数据文件,垃圾回收程序不会自动启动,你必须使用自己编写的 shell 脚本、cron 项或者其他办法来执行垃圾搜集。

比如设置 "session.gc_maxlifetime=1440"(24 分钟)。

- session.referer_check =

如果请求头中的 "Referer" 字段不包含此处指定的字符串则会话 ID 将被视为无效。注意:如果请求头中根本不存在 "Referer" 字段的话,会话 ID 仍将被视为有效。默认为空,即不做检查(全部视为有效)。

- session.entropy_file = ;"/dev/urandom"

附加的用于创建会话 ID 的外部高熵值资源(文件),例如 UNIX 系统上的 "/dev/random" 或 "/dev/urandom"。

- session.entropy_length = 0

从资源中读取的字节数(建议值:16)。

- session.use_cookies = On

是否使用 cookie 在客户端保存会话 ID。

- session.use_only_cookies = Off

是否仅仅使用 cookie 在客户端保存会话 ID,打开这个选项可以避免使用 URL 传递会话带来的安全问题。但是禁用 Cookie 的客户端将使会话无法工作。

- session.cookie_lifetime = 0

传递会话 ID 的 Cookie 有效期(秒),0 表示仅在浏览器打开期间有效。

- session.cookie_path = "/"

传递会话 ID 的 Cookie 作用路径。

- session.cookie_domain =

传递会话 ID 的 Cookie 作用域。默认为空表示根据 cookie 规范生成的主机名。

- session.cookie_secure = Off

是否仅仅通过安全连接(https)发送 cookie。

- session.cookie_httponly = Off

是否在 cookie 中添加 httpOnly 标志(仅允许 HTTP 协议访问),这将导致客户端脚本(JavaScript 等)无法访问该 cookie。打开该指令可以有效预防通过 XSS 攻击劫持会话 ID。

- session.cache_limiter = "nocache"

设为 {nocache|private|public} 以指定会话页面的缓存控制模式,或者设为空以阻止在 http 应答头中发送禁用缓存的命令。

- session.cache_expire = 180

  指定会话页面在客户端 cache 中的有效期限（分钟）session.cache_limiter＝nocache 时，此处设置无效。

- session.use_trans_sid = Off

  是否使用明码在 URL 中显示 SID（会话 ID）。默认是禁止的，因为它会给你的用户带来安全危险，用户可能将包含有效 sid 的 URL 通过 email/irc/QQ/MSN…途径告诉给其他人。包含有效 sid 的 URL 可能会被保存在公用电脑上。用户可能保存带有固定不变 sid 的 URL 在他们的收藏夹或者浏览历史纪录里面。基于 URL 的会话管理总是比基于 Cookie 的会话管理有更多的风险，所以应当禁用。

- session.bug_compat_warn = On

  "BUG"并显示警告。

- session.hash_function = 0

  生成 SID 的散列算法。SHA-1 的安全性更高一些 0：MD 5（128 bits）1：SHA-1（160 bits）建议使用 SHA-1。

- session.hash_bits_per_character = 4

  指定在 SID 字符串中的每个字符内保存多少 bit，这些二进制数是 hash 函数的运算结果 4：0—9，a—f；5：0—9，a—v；6 比特：0—9，a—z，A—Z，"—"，"，"，建议值为 5。

- url_rewriter.tags = "a=href, area=href, frame=src, form=, fieldset="

  此指令属于 PHP 核心部分，并不属于 session 模块。指定重写哪些 HTML 标签来包含 SID（仅当 session.use_trans_sid＝On 时有效）form 和 fieldset 比较特殊：如果你包含他们，URL 重写器。

以上是一些常用的 session 配置选项说明，更多的 session 配置选项说明你可以参考 php.ini 文件中的说明。

任务二　基本登录控制

创建一个登录页面，判断用户输入的账号密码是否正确。如果输入正确，提示"欢迎登录，点击此处进入欢迎界面"，如果输入错误则要求用户重新输入。

Login.html 页面代码：

```html
<html>
<head>
<title>Login</title>
<meta http-equiv="Content-Type" content="text/html; charset=gb2312">
</head>

<body>
<form name="form1" method="post" action="login.php">
  <table width="300" border="0" align="center" cellpadding="2" cellspacing="2">
    <tr>
      <td width="150"><div align="right">用户名：</div></td>
      <td width="150"><input type="text" name="username"></td>
    </tr>
    <tr>
```

```html
      <td><div align="right">密码：</div></td>
      <td><input type="password" name="passcode"></td>
    </tr>
    <tr>
      <td><div align="right">保存时间：</div></td>
      <td><select name="cookie" id="cookie">
            <option value="0" selected>浏览器进程</option>
            <option value="1">保存 1 天</option>
            <option value="2">保存 30 天</option>
            <option value="3">保存 365 天</option>
          </select></td>
    </tr>
  </table>
  <p align="center">
    <input type="submit" name="Submit" value="Submit">
    <input type="reset" name="Reset" value="Reset">
  </p>
</form>
</body>
</html>
```

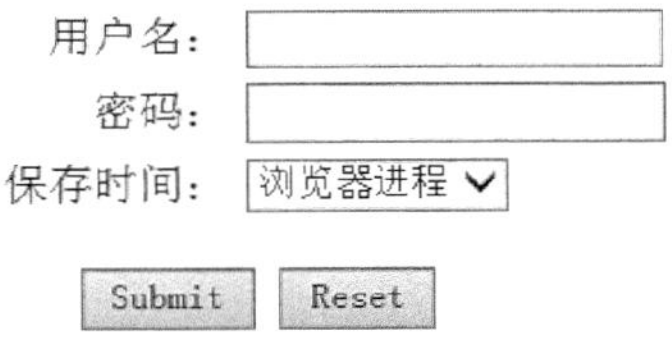

图 8.1　基本登录

Login.php 页面代码：

```php
<?php
@mysql_connect("localhost", "root","1981427")
//选择数据库之前需要先连接数据库服务器
or die("数据库服务器连接失败");
@mysql_select_db("test")                      //选择数据库 mydb
or die("数据库不存在或不可用");
//获取用户输入
$username = $_POST['username'];
$passcode = $_POST['passcode'];
//执行 SQL 语句获得 Session 的值
$query = @mysql_query("select username, userflag from users "
."where username = '$username' and passcode = '$passcode'")
or die("SQL 语句执行失败");
//判断用户是否存在,密码是否正确
if( $row = mysql_fetch_array( $query))
{
    session_start();
    //标志 Session 的开始
    //判断用户的权限信息是否有效,如果为 1 或 0 则说明有效
    if( $row['userflag'] == 1 or $row['userflag'] == 0)
```

```php
    {
        $_SESSION['username'] = $row['username'];
        $_SESSION['userflag'] = $row['userflag'];
        echo "<a href="main.php" mce_href="main.php">欢迎登录,点击此处进入欢迎界面</a>";
    }
    else
    {
        echo "用户权限信息不正确";
    }
}
else
{
    echo "用户名或密码错误";
}
?>
```

Loginout.php 页面代码:

```php
<?php
unset($_SESSION['username']);
unset($_SESSION['passcode']);
unset($_SESSION['userflag']);
echo "注销成功";
?>
```

任务三　多级权限登录控制

大多数情况下,系统的账户都会划分为不同的权限等级。比如,网络商城要分为买家账号、卖家账号、管理员账号等。要求创建登录页面,对登录的账号作出身份识别,针对不同的账号给予不同的权限、打开不同的页面或者显示不同的信息。

a.html 代码:

```html
<!DOCTYPE html PUBLIC "-//W3C//DTD XHTML 1.0 Transitional//EN"
"http://www.w3.org/TR/xhtml1/DTD/xhtml1-transitional.dtd">
<html xmlns="http://www.w3.org/1999/xhtml">
<head>
<title>a.jpg</title>
<link type="text/css" rel="stylesheet" href="css/css1.css">
<meta http-equiv="Content-Type" content="text/html; Charset="gb2312;iso-8859-1">
<meta name="description" content="FW MX CSS Layer">

</head>
<body bgcolor="#ffffff">
<div id="a">
    <div id="br1c1"><img name="b_r1_c1" src="images/b_r1_c1.jpg" width="780" height="115" border="0"></div>
    <div id="br2c1">
        <div id="cr1c1"><img name="c_r1_c1" src="images/c_r1_c1.jpg" width="139" height="335" border="0"></div>
```

```html
            <div id="cr1c2">
                <center><font color=blue size=6><br><span style="font-weight:bold;">登录用户/LOGIN</span></font><br><br>
                <form action="a.php"    method="get">
                <b><font color=blue size=5>登   录:</font></b>
                <input type="text" name="username"><br><br><br>
                <b><font color=blue size=5>密   码:</font></b>
                <input type="password" name="password"><br><br><br>
                <input type="radio" name="sf" value="学生"><font color=black size=4>学生</font>  
                <input type="radio" name="sf" value="教师"><font color=black size=4>教师</font>  
                <input type="radio"  name="sf" value="管理员"><font color=black size=4>管理员</font><br><br><br>
                <input type="submit" style=" font-size:20px;"  value="登录">    <input type="reset" style=" font-size:20px;" value="重置">
                </form>
                </center>
            </div>
            <div id="cr1c3"><img name="c_r1_c3" src="images/c_r1_c3.jpg" width="142" height="335" border="0"></div>
        </div>
        <div id="br3c1" ><img name="b_r3_c1" src="images/b_r3_c1.jpg" width="780" height="123" border="0"></div>
    </div>
    </body>
</html>
```

a.php 代码

```php
<!DOCTYPE html PUBLIC "-//W3C//DTD XHTML 1.0 Transitional//EN" "http://www.w3.org/TR/xhtml1/DTD/xhtml1-transitional.dtd">
<html xmlns="http://www.w3.org/1999/xhtml">
<head>
<meta http-equiv="Content-Type" content="text/html; charset=gb2312" />
<title>无标题文档</title>
</head>
<body>
<?php
session_start();
if(!isset($_SESSION['name'])){
    if(@$_GET['sf']==""){
    echo "您未选择身份,请选择身份!";
    header("refresh:2;url=a.htm");
        exit;
    }
    if($_GET['sf']=="学生"){
        $xh = $_GET['username'];
        $password = $_GET['password'];
        mysql_connect("127.0.0.1","root","123456");
        mysql_select_db("education");
```

```php
    $query="SELECT 学号,姓名 FROM 学生表 WHERE 学号='$xh' AND 密码='$password'";
    mysql_query("set names 'gbk'");
    $result=mysql_query($query);
    if(mysql_num_rows($result)==1){
    $_SESSION['name']=mysql_result($result,0,"姓名");
    $_SESSION['学号']=mysql_result($result,0,"学号");
        header("refresh:1;url=a2.htm");
    } else {
    echo "<script language='java script'>alert('密码错误或者未输入,请重新输入!')</script>";
        header("refresh:6;url=a.htm");
        exit;
    }
    }
    else if($_GET['sf']=="教师"){
        $jsh = $_GET['username'];
        $password = $_GET['password'];
    mysql_connect("127.0.0.1","root","123456");
    mysql_select_db("education");
        $query="SELECT 教师编号,教师姓名 FROM 教师表 WHERE 教师编号='$jsh' AND 密码='$password'";
    mysql_query("set names 'gbk'");
    $result=mysql_query($query);
    if(mysql_num_rows($result)==1){
    $_SESSION['name']=mysql_result($result,0,"教师姓名");
    $_SESSION['教师编号']=mysql_result($result,0,"教师编号");
        header("refresh:1;url=a3.htm");
    } else {
        echo "密码错误或者未输入,请重新输入!";
        header("refresh:2;url=a.htm");
        exit;
    }
    }
    else if($_GET['sf']=="管理员"){
        $ab = $_GET['username'];
        $password = $_GET['password'];
        mysql_connect("127.0.0.1","root","123456");
        mysql_select_db("education");
            $query="SELECT 管理员号,管理员 FROM 管理员表 WHERE 管理员号='$ab' AND 密码='$password'";
    mysql_query("set names 'gbk'");
    $result=mysql_query($query);
    if(mysql_num_rows($result)==1){
    $_SESSION['name']=mysql_result($result,0,"管理员");
    $_SESSION['管理员号']=mysql_result($result,0,"管理员号");
        header("refresh:1;url=a4.htm");
    } else {
        echo "密码错误或者未输入,请重新输入!";
        header("refresh:2;url=a.htm");
        exit;
    }
    }
```

        }
?>
</body>
</html>

代码运行效果如图 8.2 所示。

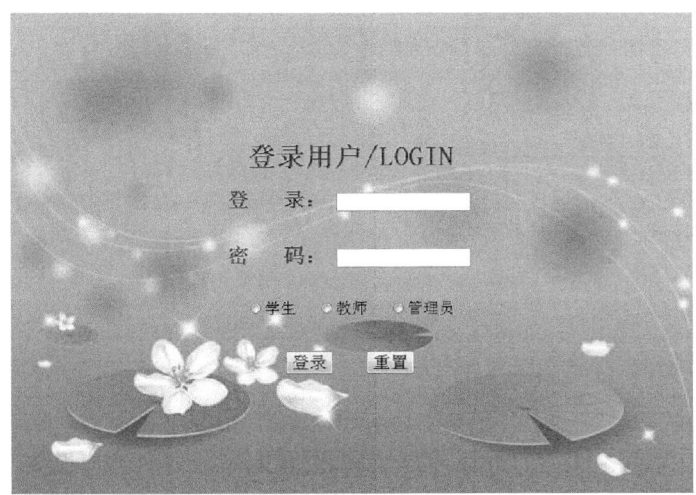

图 8.2　多身份登录

任务四　带验证码的登录控制

为了防止黑客的恶意攻击,通常会采用验证码的形式来限制黑客程序的自动登录。因为,程序是很难识别图形中包含的验证码的,而人眼做到这一点确没有一点问题。要求创建验证码生成页面,在登录页面调用生成验证码。

checkcode.php 代码如下:

```
<?
session_start();
for($i=0; $i<4; $i++){
$rand. = dechex(rand(1,15));
}
$_SESSION[check_pic]= $rand;
//echo $_SESSION[check_pic];
// 设置图片大小
$im = imagecreatetruecolor(100,30);
// 设置颜色
$bg = imagecolorallocate($im,0,0,0);
$te = imagecolorallocate($im,255,255,255);
// 把字符串写在图像左上角
imagestring($im,rand(5,6),rand(25,30),5,$rand,$te);
// 输出图像
header("Content-type:image/jpeg");
imagejpeg($im);
?>
```

form.php 页面:

通过 `<img src="checkcode.php">` 调用生成的验证码图片，代码如下:

```html
<div class="bottomAds">
<fieldset class="bottomAds_quote"><legend>留言</legend>
<div class="ads">
<form action="../utity/post.php" method="post" onsubmit="return chkinput(this)">
<input name="name" type="text" />您的名字
<input name="email" type="text" />您的邮件
<input name="website" type="text" />您的网站
<textarea name="content" style="width:340; height:150;">
</textarea><br />
<img src="checkcode.php"><input type="text" name="check"><br />
<input type="submit" value="提交" />
</form>
</div>
<br clear="both" />
</fieldset>
```

post.php 页面:

比较 $_POST[check] 与 $_SESSION[check_pic],若相等则执行数据库插入操作。不相等就返回上一页。代码如下:

```php
<?php
session_start();
if(isset($_POST[check]))
{
if($_POST[check] == $_SESSION[check_pic])
{
// echo "验证码正确".$_SESSION[check_pic];
require("dbinfo.php");
$name = $_POST['name'];
$email = $_POST['email'];
$website = $_POST['website'];
$content = $_POST['content'];
$date = date("Y-m-d h:m:s");
// 连接到 MySQL 服务器
$connection = mysql_connect($host, $username, $password);
if (!$connection)
{
die('Not connected : ' . mysql_error());
}
// 设置活动的 MySQL 数据库
$db_selected = mysql_select_db($database, $connection);
if (!$db_selected)
{
die ('Can\'t use db : ' . mysql_error());
}
// 向数据库插入数据
$query = "insert into table (nowamagic_name, nowamagic_email, nowamagic_website, nowamagic_content, nowamagic_date) values ('$name','$email','$website','$content','$date')";
```

```
$result = mysql_query($query);
if($result)
{
echo "<script>alert('提交成功');history.go(-1);</script>";
}
if (!$result)
{
die('Invalid query: ' . mysql_error());
}
}
else
{
echo "<script>alert('验证码错误');history.go(-1);</script>";
}
}
?>
```

五、考核标准

（1）版面布局合理清晰，整体效果美观，观赏性强。（20分）

（2）网页中没有明显的错误（如超链接、图片无法显示、错别字等）。（10分）

（3）完成 PHP SESSION 配置。（10分）

（4）完成账号密码验证。（20分）

（5）完成多用户登录认证。（30分）

项目九 FCKeditor 应用

一、实训目的

- 掌握 FCKeditor 的基本用法；
- 能够在 FCKeditor 的基础上，进行功能扩展。

二、实训要求

- 下载 FCKeditor，正确添加到项目中；
- 对 FCKeditor 进行配置，能够精确控制其中的各种参数，实现不同的效果；
- 熟练使用 FCKeditor 进行各种新闻、公告、通知的发布和管理。

三、实训设计

新闻公告是每一个网站必不可少的内容，广泛出现在各类大中小网站之中。FCKeditor 是一个专门使用在网页上属于开放源代码的所见即所得文字编辑器。它志于轻量化，不需要太复杂的安装步骤即可使用。本实训通过 FCKeditor 进行新闻编辑，需要学生在实训过程中熟悉 FCKeditor 关于 PHP 部分的使用方法，熟练的创建新闻编辑页面、新闻存储页面、新闻浏览页面。

四、实训内容

任务一 FCKeditor 的下载和配置

（1）下载并解压缩 FCKeditor 后，打开 FCKeditor 目录下的 _samples 文件夹，里面含有各种编程语言调用 FCKeditor 的范例程序页面，其中 PHP 目录中包含着一些使用 PHP 来调用 FCKeditor 的范例。

（2）工具栏设置。默认情况下，FCKeditor 会调用如下的工具栏按钮，大家可以根据自己的需要进行增减。需要注意的是，2.0 版与 1.6 版的按钮并不完全相同，有些按钮以及删除或者改名了。

```
//##
//## Toolbar Buttons Sets
//##
FCKConfig.ToolbarSets["Default"] = [
    ['Source','-','Save','NewPage','Preview'],
    ['Cut','Copy','Paste','PasteText','PasteWord','-','Print'],
    ['Undo','Redo','-','Find','Replace','-','SelectAll','RemoveFormat'],
['Bold','Italic','Underline','StrikeThrough','-','Subscript','Superscript'],
    ['OrderedList','UnorderedList','-','Outdent','Indent'],
    ['JustifyLeft','JustifyCenter','JustifyRight','JustifyFull'],
    ['Link','Unlink'],
    ['Image','Table','Rule','SpecialChar','Smiley'],
    ['Style','FontFormat','FontName','FontSize'],
```

```
        ['TextColor','BGColor'],
        ['About']
];
```

（3）简体中文设置，编辑 edit/lang/fcklanguagemanager.js，将下面语句：

```
FCKLanguageManager.AvailableLanguages =
{
'ar'    : 'Arabic',
'bs'    : 'Bosnian',
'ca'    : 'Catalan',
'en'    : 'English',
'es'    : 'Spanish',
'et'    : 'Estonian',
'fi'    : 'Finnish',
'fr'    : 'French',
'gr'    : 'Greek',
'he'    : 'Hebrew',
'hr'    : 'Croatian',
'it'    : 'Italian',
'ko'    : 'Korean',
'lt'    : 'Lithuanian',
'no'    : 'Norwegian',
'pl'    : 'Polish',
'sr'    : 'Serbian (Cyrillic)',
'sr-latn' : 'Serbian (Latin)',
'sv'    : 'Swedish'
}
```

添加一行 'zh-cn'：'Chinese' 从而变成：

```
FCKLanguageManager.AvailableLanguages =
{
'ar'    : 'Arabic',
'bs'    : 'Bosnian',
'ca'    : 'Catalan',
'en'    : 'English',
'es'    : 'Spanish',
'et'    : 'Estonian',
'fi'    : 'Finnish',
'fr'    : 'French',
'gr'    : 'Greek',
'he'    : 'Hebrew',
'hr'    : 'Croatian',
'it'    : 'Italian',
'ko'    : 'Korean',
'lt'    : 'Lithuanian',
'no'    : 'Norwegian',
'pl'    : 'Polish',
'sr'    : 'Serbian (Cyrillic)',
'sr-latn' : 'Serbian (Latin)',
'sv'    : 'Swedish',
```

```
'zh-cn'        : 'Chinese'
}
```

然后从 http://www.shaof.com/download/zh-cn.js 下载汉化好的 zh-cn.js 保存到 editor/lang 目录下即可。

（4）修改上传路径。FCKeditor 2.0 在线编辑器采用了一种名为"Connector"（连接器）的技术来实现对文件的浏览以及上传。首先修改配置文件 FCKeditor/fckconfig.js 中的两段内容：

```
//Link Browsing
FCKConfig.LinkBrowser = true ;
FCKConfig.LinkBrowserURL = FCKConfig.BasePath +
"filemanager/browser/default/browser.html?Connector=connectors/php/connector.php" ;
FCKConfig.LinkBrowserWindowWidth  = screen.width * 0.7 ;    // 70%
FCKConfig.LinkBrowserWindowHeight = screen.height * 0.7 ;   // 70%
//Image Browsing
FCKConfig.ImageBrowser = true ;
FCKConfig.ImageBrowserURL = FCKConfig.BasePath +
" filemanager/browser/default/browser. html? Type = Image&Connector = connectors/php/
connector.php" ;
FCKConfig.ImageBrowserWindowWidth  = screen.width * 0.7 ;    // 70% ;
FCKConfig.ImageBrowserWindowHeight = screen.height * 0.7 ;   // 70% ;
```

然后修改配置文件。

```
FCKeditor/editor/filemanager/browser/default/connectors/php/connector.php
// Get the "UserFiles" path.
$GLOBALS["UserFilesPath"] = '/UserFiles/' ;
```

UserFiles 为文件上传的路径，与本文开头所给的例子相对应，大家可以自行修改。

（5）修改上传文件类型。Fckeditor2.6.6 默认并没有限制上传图片文件的大小，可以通过两种方法改进：一种可以通过修改 PHP.INI 配置文件上传大小来限制，另一种方法只能手动修改 Fckeditor 源码，首先打开 editor/filemanager/connectors/php 目录下 config.php，创建 Config 变量设置上传图片大小，这里以 KB 为单位。复制代码，代码如下：

```
$Config['MaxImageSize']= '1024';
```

其次打开 editor/filemanager/connectors/php 目录下 commands.php，复制代码，代码如下：

```
if( isset( $Config['SecureImageUploads'] ) )
{
if( ( $isImageValid = IsImageValid( $oFile['tmp_name'], $sExtension ) ) === false )
{
$sErrorNumber = '202' ;
}
//上传图片大小限制
}
```

在上传图片大小限制处，添加代码，代码如下：

```
if( isset( $Config['MaxImageSize'] ) )
{
    $iFileSize = round( $oFile['size'] / 1024 );
    if( $iFileSize > $Config['MaxImageSize'] )
```

```
        {
            $sErrorNumber = '204';
        }
}
```

说明：由于 PHP 计算上传图片大小以字节为单位，所以代码首先将上传的图片大小折算为 KB，再来对比是否超出了规定的图片大小，如超出，则报错。代码如下：

```
if( ! $sErrorNumber && IsAllowedExt( $sExtension, $resourceType ) )
{
//Fckeditor 上传图片功能
}
else
$sErrorNumber = '202';
```

代码块结尾处的 else 语句去除，否则实现不了限制 Fckeditor 上传图片文件大小的功能。

最后打开 editor/dialog/fck_image/fck_image.js，添加错误代码（errorNumber）信息，找到 OnUploadCompleted 函数，添加代码如下：

```
case 204 :
alert( "Security error. File size error." );
return ;
```

至此限制 Fckeditor 上传图片文件大小配置就完成了，其他类型的上传文件大小限制也是这种思路。

（6）修改上传文件类型。找到 FCKeditor\editor\filemanager\connectors\php\ 文件夹下的 config.php 文件，找到下面的这一行！

```
ConfigAllowedExtensions.Add "File",
"7z|aiff|asf|avi|bmp|csv|doc|fla|flv|gif|gz|gzip|jpeg|jpg|mid|mov|mp3|mp4|mpc|mpeg|mpg|ods|odt|pdf|png|ppt|pxd|qt|ram|rar|rm|rmi|rmvb|rtf|sdc|sitd|swf|sxc|sxw|tar|tgz|tif|tiff|txt|vsd|wav|wma|wmv|xls|xml|zip"
```

加入你要的扩展名，当然这里有"File, Image, Flash, Media" 4 个类别设置，随你选哪一个加进去就可以了！注意和其他的扩展名用"|"符号分开。

任务二　新闻编辑

完成 FCKeditor 的下载配置后，我们开始使用 FCKeditor 进行新闻编辑。

第一次进行新闻发布，我们可以使用测试目录下的新闻发布页面，在测试页面的基础上进行修改，完成新闻发布功能。

（1）打开 fckeditor_samples\php 下的 sample01.php，并进行配置，代码如下：

```
<! DOCTYPE HTML PUBLIC "-//W3C//DTD HTML 4.0 Transitional//EN">
<html>
    <head>
        <title>FCKeditor - Sample</title>
        <meta http-equiv="Content-Type" content="text/html; charset=utf-8">
        <meta name="robots" content="noindex, nofollow">
        <link href="../sample.css" rel="stylesheet" type="text/css" />
```

```
            </head>
            <body>
                <h1>FCKeditor - PHP - Sample 1</h1>
                This sample displays a normal HTML form with an FCKeditor with full features
                enabled.
                <hr>
                <form action="samplepostdata.php" method="post" target="_blank">
<?php
// Automatically calculates the editor base path based on the _samples directory.
// This is usefull only for these samples. A real application should use something like this:
// $oFCKeditor->BasePath = '/fckeditor/';    // '/fckeditor/' is the default value.

$sBasePath = $_SERVER['PHP_SELF'];
$sBasePath = substr( $sBasePath, 0, strpos( $sBasePath, "_samples" ) );
$oFCKeditor = new FCKeditor('FCKeditor1') ;
$oFCKeditor->Width  = '90%';
$oFCKeditor->Height = '300px';
$oFCKeditor->BasePath = $sBasePath ;
$oFCKeditor->Value    = '<p>This is some <strong>sample text</strong>. You are using <a href="http://www.fckeditor.net/">FCKeditor</a>.</p>';
$oFCKeditor->Create() ;
?>
                <br>
                <input type="submit" value="Submit">
            </form>
        </body>
</html>
```

其中 $oFCKeditor 代表这个新创建的编辑器对象名称，$oFCKeditor->Width 控制编辑器的宽度，$oFCKeditor->Height 控制编辑器的高度，$oFCKeditor->Value 控制编辑器打开后显示的内容。action="samplepostdata.php"指明了新闻提交的服务器页面。效果如图 9.1 所示。

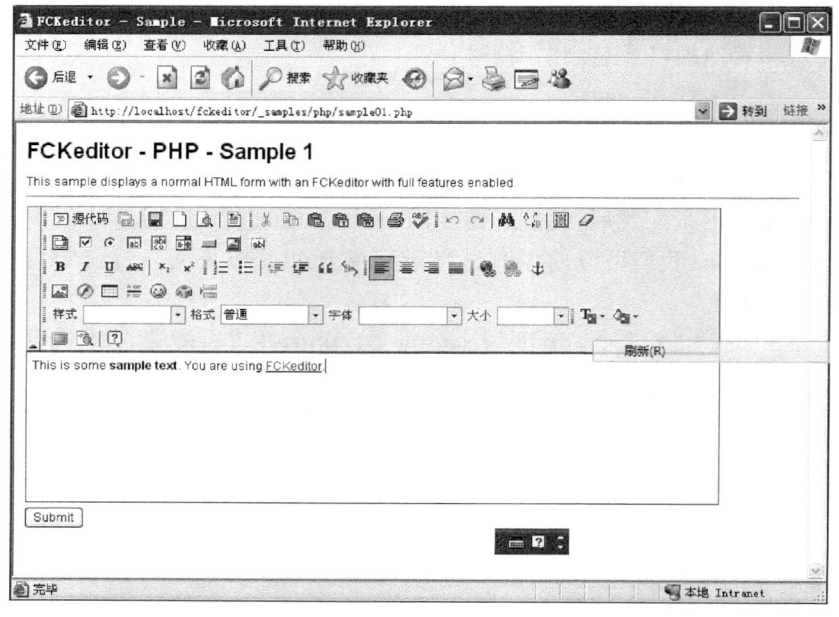

图 9.1　新闻编辑页面

在本页面进行新闻编辑后,点击"submit"按钮,将新闻提交至服务器,由"sampleposteddata.php"页面负责对数据进行处理。

(2) 本页面接收新闻编辑页面的数据后,进行处理并存入数据库,具体代码如下:

```php
<? php
/*
* FCKeditor - The text editor for Internet - http://www.fckeditor.net
* Copyright (C) 2003-2010 Frederico Caldeira Knabben
*
* == BEGIN LICENSE ==
*
* Licensed under the terms of any of the following licenses at your
* choice:
*
*  - GNU General Public License Version 2 or later (the "GPL")
*    http://www.gnu.org/licenses/gpl.html
*
*  - GNU Lesser General Public License Version 2.1 or later (the "LGPL")
*    http://www.gnu.org/licenses/lgpl.html
*
*  - Mozilla Public License Version 1.1 or later (the "MPL")
*    http://www.mozilla.org/MPL/MPL-1.1.html
*
* == END LICENSE ==
*
* This page lists the data posted by a form.
*/
?>
<!DOCTYPE HTML PUBLIC "-//W3C//DTD HTML 4.0 Transitional//EN">
<html>
    <head>
        <title>FCKeditor - Samples - Posted Data</title>
        <meta http-equiv="Content-Type" content="text/html; charset=utf-8">
        <meta name="robots" content="noindex, nofollow">
        <link href="../sample.css" rel="stylesheet" type="text/css">
    </head>
    <body>
        <h1>FCKeditor - Samples - Posted Data</h1>
        This page lists all data posted by the form.
        <hr>
        <table border="1" cellspacing="0" id="outputSample">
            <colgroup><col width="80"><col></colgroup>
            <thead>
                <tr>
```

```
                <th>Field Name</th>
                <th>Value</th>
            </tr>
        </thead>
<?php

if ( isset( $_POST ) )
    $postArray = &$_POST ;           // 4.1.0 or later, use $_POST
else
    $postArray = &$HTTP_POST_VARS ;  // prior to 4.1.0, use HTTP_POST_VARS

foreach ( $postArray as $sForm => $value )
{
    if ( get_magic_quotes_gpc() )
        $postedValue = htmlspecialchars( stripslashes( $value ) ) ;
    else
        $postedValue = addslashes( $value ) ;

?>
            <tr>
                <th><?php echo htmlspecialchars( $sForm ) ?></th>
                <td><pre><?php echo $postedValue ?></pre></td>
            </tr>
<?php
}
$link=mysql_connect("localhost","root","majie")
    or die("数据库服务器连接失败！<BR>");
mysql_select_db("test",$link) or die("数据库选择失败！<BR>");
mysql_query("set names 'utf8'");
$sql="INSERT INTO 'news' VALUES ('新闻测试','小王','新部门','$postedValue')";
if (mysql_query($sql,$link))
    echo "增加成功!";
else
    echo '增加失败！';
?>
        </table>
    </body>
</html>
```

要注意的是，本次实训中，为了便于测试，向数据库写入数据时，除了新闻内容，都为固定值。请同学们在测试完成后，自行修改新闻编辑页面，增加相关信息，将固定值改为变量。

任务三　新闻浏览页面

（1）新闻正确存入数据库后，就可以在网页上打开浏览，首先打开"editor.php"页面，可以看

到所有的新闻标题,具体代码如下:

```php
<head>
<meta http-equiv="Content-Type" content="text/html; charset=gb2312" />
<title>无标题文档</title>
</head>

<body>
<?php

$link=mysql_connect("localhost","root","majie")
    or die("数据库服务器连接失败!<BR>");
  mysql_select_db("test",$link) or die("数据库选择失败!<BR>");
  mysql_query("set names 'gbk'");
  $sql="select * from student ";
  $result=mysql_query($sql,$link);
  $row = mysql_fetch_array($result);
  mysql_data_seek($result,0);
  while($row = mysql_fetch_array($result))
  {
  ?>
  <a href="text.php?name=<?php echo $row['name']; ?>" target="_blank"><?php echo $row['name']; ?></a> <br />
  <?php
  }
  ?>
</body>
</html>
```

(2) 点击标题后,就可以看到具体的新闻,代码如下:

```php
<head>
<meta http-equiv="Content-Type" content="text/html; charset=gb2312" />
<title>无标题文档</title>
</head>
<body>
<?php
$name=$_GET['name'];
$link=mysql_connect("localhost","root","majie")
    or die("数据库服务器连接失败!<BR>");
  mysql_select_db("test",$link) or die("数据库选择失败!<BR>");
  mysql_query("set names 'gbk'");
  $sql="select * from student where name='$name'";
  $result=mysql_query($sql,$link);
  $row = mysql_fetch_array($result);
  echo $row['stext'];
```

```
//stripslashes
//addslashes
 ?>
</body>
</html>
```

代码运行结果如图 9.2 所示。

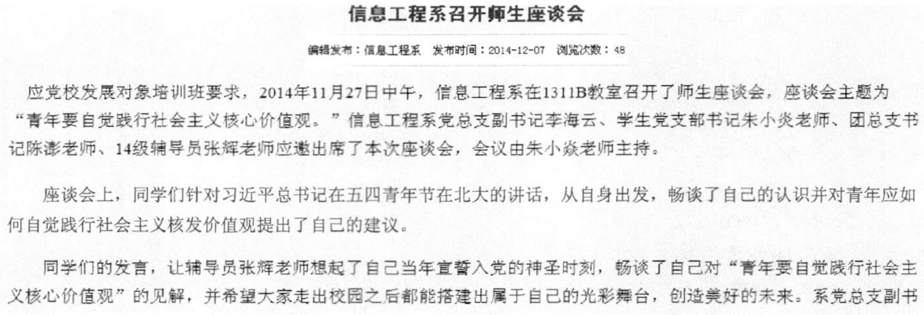

信息工程系召开师生座谈会

编辑发布：信息工程系　发布时间：2014-12-07　浏览次数：48

应党校发展对象培训班要求，2014年11月27日中午，信息工程系在1311B教室召开了师生座谈会，座谈会主题为"青年要自觉践行社会主义核心价值观。"信息工程系党总支副书记李海云、学生党支部书记朱小炎老师、团总支书记陈澎老师、14级辅导员张辉老师应邀出席了本次座谈会，会议由朱小焱老师主持。

座谈会上，同学们针对习近平总书记在五四青年节在北大的讲话，从自身出发，畅谈了自己的认识并对青年应如何自觉践行社会主义核发价值观提出了自己的建议。

同学们的发言，让辅导员张辉老师想起了自己当年宣誓入党的神圣时刻，畅谈了自己对"青年要自觉践行社会主义核心价值观"的见解，并希望大家走出校园之后都能搭建出属于自己的光彩舞台，创造美好的未来。系党总支副书

图 9.2　新闻浏览页面

五、考核标准

（1）版面布局合理清晰，整体效果美观，观赏性强。（10 分）

（2）网页中没有明显的错误（如超链接、图片无法显示、错别字等）。（10 分）

（3）安装配置 FCKeditor。（20 分）

（4）使用 FCKeditor 创建新闻编辑和存储页面。（30 分）

（5）使用 FCKeditor 创建新闻浏览页面。（30 分）

项目十 数据导入 Office

一、实训目的

- 掌握网页数据导入 Excel 的基本用法；
- 掌握网页数据导入 Word 的基本用法。

二、实训要求

- 将网页数据导入 Excel，并编辑打印；
- 将网页数据导入 Word，并编辑打印。

三、实训设计

管理类的系统中，经常见到各种表格，很多时候需要将这些表格中的数据导出，并进行编辑后打印、签字和盖章。日常工作中，除了导入 Excel 外，导入 Word 也是常用到的功能。本实训通过 PHP 将数据库中的数据显示在网页之中，再通过 JS 代码将页面中的数据导入 Excel 和 Word 之中。

四、实训内容

任务一 数据导入 Excel

将数据导入 Excel 的方式主要有两种：第一种相对简单，如果网页中的表格内容和样式在导出后不需要进行修改，则可以选择这种方式：

从数据库中读取数据，在页面生成表格，并将表格内的数据导入 Excel。

方法 1：

```
<SCRIPT LANGUAGE = "javascript">
function method1(tableid) {//整个表格拷贝到 EXCEL 中
    var curTbl = document.getElementById(tableid);
    var oXL = new ActiveXObject("Excel.Application");
    //创建 AX 对象 excel
    var oWB = oXL.Workbooks.Add();
    //获取 workbook 对象
    var oSheet = oWB.ActiveSheet;
    //激活当前 sheet
    var sel = document.body.createTextRange();
    sel.moveToElementText(curTbl);
    //把表格中的内容移到 TextRange 中
    sel.select();
    //全选 TextRange 中内容
    sel.execCommand("Copy");
    //复制 TextRange 中内容
    oSheet.Paste();
```

```
        //粘贴到活动的 EXCEL 中
        oXL.Visible = true;
        //设置 excel 可见属性
    }
    </SCRIPT>
```

此种方法不用关心网页中表格显示的内容,是全复制的导入,只要将

```
var curTbl = document.getElementById(tableid)
```

中的"tableid"替换为自己页面中的表格 ID 即可。

不过,大多数时候,数据的导出,会面临复杂表格的情况,下面我们来看下复杂表格的导出代码。

方法 2:

```
<html>
<style type="text/css">
#customers
    {
    font-family:"Trebuchet MS", Arial, Helvetica, sans-serif;
    width:130%;
    border-collapse:collapse;
    }

#customers td, #customers th
    {
    font-size:0.8em;
    border:1px solid #98bf21;
    
    
    }

#customers th
    {
    font-size:0、8em;
    text-align:center;
    padding-top:5px;
    padding-bottom:4px;
    background-color:#A7C942;
    color:#ffffff;
    }

#customers tr.alt td
    {
    color:#000000;
    background-color:#EAF2D3;
    }
</style>
<SCRIPT LANGUAGE="JavaScript">
<!--
    function printt()//导出 excel
    {
    var oXL = new ActiveXObject("Excel.Application");
```

```javascript
var oWB = oXL.Workbooks.Add();
var oSheet = oWB.Worksheets(1);
var Lenr = customers.rows.length;
oSheet.Range("A3","AF25").Borders.Weight=2;//单元格边框
oSheet.Range("A1","AF25").HorizontalAlignment=3;//内容居中显示
oSheet.Range("A2","G2").HorizontalAlignment=2;//内容居左显示
//设置字体
oSheet.Range("A1","AF1").Font.Size=20
//在此进行样式控制
    oSheet.Rows(1+":"+1).RowHeight = 28;//定义行高
    oSheet.Rows(2+":"+2).RowHeight = 28;
    //定义列宽
    oSheet.Columns('A:A').ColumnWidth = 5;
    oSheet.Columns('B:B').ColumnWidth = 5;
    oSheet.Columns('C:C').ColumnWidth = 10;
    oSheet.Columns('D:D').ColumnWidth = 4;
    oSheet.Columns('E:E').ColumnWidth = 4;
    oSheet.Columns('F:F').ColumnWidth = 4;
    oSheet.Columns('G:G').ColumnWidth = 4;
    oSheet.Columns('H:H').ColumnWidth = 4;
    oSheet.Columns('I:I').ColumnWidth = 4;
    oSheet.Columns('J:J').ColumnWidth = 4;
    oSheet.Columns('K:K').ColumnWidth = 4;
    oSheet.Columns('L:L').ColumnWidth = 4;
    oSheet.Columns('M:M').ColumnWidth = 4;
    oSheet.Columns('N:N').ColumnWidth = 4;
    oSheet.Columns('O:O').ColumnWidth = 4;
    oSheet.Columns('P:P').ColumnWidth = 4;
    oSheet.Columns('Q:Q').ColumnWidth = 4;
    oSheet.Columns('R:R').ColumnWidth = 4;
    oSheet.Columns('S:S').ColumnWidth = 4;
    oSheet.Columns('T:T').ColumnWidth = 4;
    oSheet.Columns('U:U').ColumnWidth = 4;
    oSheet.Columns('V:V').ColumnWidth = 4;
    oSheet.Columns('W:W').ColumnWidth = 4;
    oSheet.Columns('X:X').ColumnWidth = 4;
    oSheet.Columns('Y:Y').ColumnWidth = 4;
    oSheet.Columns('Z:Z').ColumnWidth = 4;
    oSheet.Columns('AA:AA').ColumnWidth = 4;
    oSheet.Columns('AB:AB').ColumnWidth = 4;
    oSheet.Columns('AC:AC').ColumnWidth = 4;
    oSheet.Columns('AD:AD').ColumnWidth = 4;
    oSheet.Columns('AE:AE').ColumnWidth = 4;
    oSheet.Columns('AF:AF').ColumnWidth = 8;
    //自动换行
    oSheet.Range("C3","C4").WrapText=true;
    //合并单元格
  oSheet.Range("A1","AF1").mergecells=true;
  oSheet.Range("A2","G2").mergecells=true;
  oSheet.Range("Z2","AE2").mergecells=true;
  oSheet.Range("A3","A4").mergecells=true;
```

```
oSheet.Range("A3","A4").mergecells=true;
oSheet.Range("B3","B4").mergecells=true;
oSheet.Range("C3","C4").mergecells=true;
oSheet.Range("D3","E3").mergecells=true;
oSheet.Range("F3","G3").mergecells=true;
oSheet.Range("H3","I3").mergecells=true;
oSheet.Range("J3","K3").mergecells=true;
oSheet.Range("L3","M3").mergecells=true;
oSheet.Range("N3","O3").mergecells=true;
oSheet.Range("P3","Q3").mergecells=true;
oSheet.Range("R3","S3").mergecells=true;
oSheet.Range("T3","U3").mergecells=true;
oSheet.Range("V3","W3").mergecells=true;
oSheet.Range("X3","Y3").mergecells=true;
oSheet.Range("Z3","AA3").mergecells=true;
oSheet.Range("AB3","AC3").mergecells=true;
oSheet.Range("AD3","AE3").mergecells=true;
oSheet.Range("AF3","AF4").mergecells=true;

for(i=0;i<Lenr;i++)
{
  var Lenc = customers.rows(i).cells.length;

  for(j=0;j<Lenc-1;j++)
  {
   oSheet.Columns(2).NumberFormatLocal="@";
   oSheet.Cells(i+3,j+1).value = customers.rows(i).cells(j).innerText;
  }

}
//oSheet.PageSetup.TopMargin=4/0.035;
//单元格内容
oXL.Visible = true;
oSheet.Range("A1","AF1").value="慧龙公司医疗保险统计表";
oSheet.Range("A2","G2").value="单位:(元)";
oSheet.Range("Z2","AE2").value="年      月      日";
oSheet.Range("F3","F3").value="1 月";
oSheet.Range("H3","I3").value="2 月";
oSheet.Range("J3","K3").value="3 月";
oSheet.Range("L3","M3").value="4 月";
oSheet.Range("N3","O3").value="5 月";
oSheet.Range("P3","Q3").value="6 月";
oSheet.Range("R3","S3").value="7 月";
oSheet.Range("T3","U3").value="8 月";
oSheet.Range("V3","W3").value="9 月";
oSheet.Range("X3","Y3").value="10 月";;
oSheet.Range("Z3","AA3").value="11 月";
oSheet.Range("AB3","AC3").value="12 月";
oSheet.Range("AD3","AE3").value="本年合计";
oSheet.Range("D4","D4").value="单位";
oSheet.Range("E4","E4").value="个人";
```

```
oSheet.Range("F4","F4").value="单位";
oSheet.Range("G4","G4").value="个人";
oSheet.Range("H4","H4").value="单位";
oSheet.Range("I4","I4").value="个人";
oSheet.Range("J4","J4").value="单位";
oSheet.Range("K4","K4").value="个人";
oSheet.Range("L4","L4").value="单位";
oSheet.Range("M4","M4").value="个人";
oSheet.Range("N4","N4").value="单位";
oSheet.Range("O4","O4").value="个人";
oSheet.Range("P4","P4").value="单位";
oSheet.Range("Q4","Q4").value="个人";
oSheet.Range("R4","R4").value="单位";
oSheet.Range("S4","S4").value="个人";
oSheet.Range("T4","T4").value="单位";
oSheet.Range("U4","U4").value="个人";
oSheet.Range("V4","V4").value="单位";
oSheet.Range("W4","W4").value="个人";
oSheet.Range("X4","X4").value="单位";
oSheet.Range("Y4","Y4").value="个人";
oSheet.Range("Z4","Z4").value="单位";
oSheet.Range("AA4","AA4").value="个人";
oSheet.Range("AB4","AB4").value="单位";
oSheet.Range("AC4","AC4").value="个人";
oSheet.Range("AD4","AD4").value="单位";
oSheet.Range("AE4","AE4").value="个人";
oSheet.Range("AD3","AE3").value="本年合计";
oSheet.Range("AF3","AF4").value="本年累计";
// oSheet.Range("A1","AF2").mergecells=true;
//    oSheet.Range("A1","AF2").value="本年累计";
   }
//-->
</SCRIPT>
<body>
<? php
error_reporting(0);
 $pgroup=$_GET["pgroup"];
   $work=$_GET["work"];
    require("conn.php");
?>
<form action="yb.php" method="get">
<select name="pgroup" size="1">
<option value="<? php echo $pgroup ? >" ><? php echo $pgroup ? ></option>
<option value="2011">2011</option>
<option value="2010">2010</option>
<option value="2009">2009</option>
<option value="2008">2008</option>
<option value="2007">2007</option>
<option value="2006">2006</option>
<option value="2005">2005</option>
<option value="2004">2004</option>
```

```
<option value="2003">2003</option>
<option value="2002">2002</option>
<option value="2001">2001</option>
<option value="2000">2000</option>
</select>
  <input name="submit" type="submit" value="选择">
</form>
<INPUT TYPE="button" id="print" value="打印" name="print" onClick="printt()">

<?php
/* select id,kid,work,name,sex,idcard,contrst,contred,gotime,kind,city,phone,note from contr  where kid='1' order by id; */
    $sql="select * from yb where yeadate='$pgroup' ";
    $result=$link->query($sql);
    $rows=$result->num_rows;   //总记录数
    if($rows==0) {
      echo "没有满足条件的记录!";
      die();
    }
    $pagesize=10;  //每页的记录数(在此暂设为5,通常应设为10)
    $pagecount=ceil($rows/$pagesize);   //总页数
    //$pageno 的值为当前页的页号
    if(!isset($pageno)||$pageno<1)
       $pageno=1;
    if($pageno>$pagecount)
       $pageno=$pagecount;
    $offset=($pageno-1)*$pagesize;
    $result->data_seek($offset);
?>
<h4>慧龙公司医疗保险统计表:</h4>
<table border="1" id="customers" align="center">
<tr>
    <th rowspan="2">序号</th>
    <th rowspan="2">姓名</th>
    <th rowspan="2" width="80px">上年接转累计金额</th>
    <th colspan="2">上年接转</th>
    <th colspan="2">1月</th>
    <th colspan="2">2月</th>
    <th colspan="2">3月</th>
    <th colspan="2">4月</th>
    <th colspan="2">5月</th>
    <th colspan="2">6月</th>
    <th colspan="2">7月</th>
    <th colspan="2">8月</th>
    <th colspan="2">9月</th>
    <th colspan="2">10月</th>
    <th colspan="2">11月</th>
    <th colspan="2">12月</th>
    <th colspan="2">本年合计</th>
    <th rowspan="2">本年累计</th>
```

```
          <th rowspan="2">操作</th>
      </tr>
      <tr>
          <th>单位</th>
          <th>个人</th>
          <th>单位</th>
          <th>个人</th>
          <th>单位</th>
          <th>个人</th>
          <th>单位</th>
          <th>个人</th>
          <th>单位</th>
          <th>个人</th>
          <th>单位</th>
          <th>个人</th>
          <th>单位</th>
          <th>个人</th>
          <th>单位</th>
          <th>个人</th>
          <th>单位</th>
          <th>个人</th>
          <th>单位</th>
          <th>个人</th>
          <th>单位</th>
          <th>个人</th>
          <th>单位</th>
          <th>个人</th>
      </tr>
      <?php
         $i=0;
         while($row=$result->fetch_object()) {
         $nename=$row->name;
         $yedate=$row->yeadate-1;
         $newsql="select * from yb where name='$nename' and yeadate=$yedate ";//求上年总数
         $newresult=$link->query($newsql);
         $newrow=$newresult->fetch_object();
         $allsql="select sum(bmoney)+sum(amoney) as allmoney from yb where name='$nename' and yeadate<=$yedate+1 ";//求至今为止的总数
         $allresult=$link->query($allsql);
         $allrow=$allresult->fetch_object();
         //计算合计
         $hjold=$hjold+$newrow->amoney+$newrow->bmoney;
         $hjaold=$hjaold+$newrow->amoney;
         $hjbold=$hjbold+$newrow->bmoney;
         $hja1=$hja1+$row->a1date;
         $hjb1=$hjb1+$row->b1date;
         $hja2=$hja2+$row->a2date;
```

```php
    $hjb2 = $hjb2 + $row->b2date;
    $hja3 = $hja3 + $row->a3date;
    $hjb3 = $hjb3 + $row->b3date;
    $hja4 = $hja4 + $row->a4date;
    $hjb4 = $hjb4 + $row->b4date;
    $hja5 = $hja5 + $row->a5date;
    $hjb5 = $hjb5 + $row->b5date;
    $hja6 = $hja6 + $row->a6date;
    $hjb6 = $hjb6 + $row->b6date;
    $hja7 = $hja7 + $row->a7date;
    $hjb7 = $hjb7 + $row->b7date;
    $hja8 = $hja8 + $row->a8date;
    $hjb8 = $hjb8 + $row->b8date;
    $hja9 = $hja9 + $row->a9date;
    $hjb9 = $hjb9 + $row->b9date;
    $hja10 = $hja10 + $row->a10date;
    $hjb10 = $hjb10 + $row->b10date;
    $hja11 = $hja11 + $row->a11date;
    $hjb11 = $hjb11 + $row->b11date;
    $hja12 = $hja12 + $row->a12date;
    $hjb12 = $hjb12 + $row->b12date;
    $hjamoney = $hjamoney + $row->amoney;
    $hjbmoney = $hjbmoney + $row->bmoney;
    $hjallmoney = $hjallmoney + $allrow->allmoney;
?>
<tr>
<td><div align="center"><?php echo $row->ybid; ?></div></td>
  <td><div align="center"><?php echo $row->name; ?></div></td>
    <td><div align="center"><?php echo
$newrow->amoney + $newrow->bmoney; ?></div></td>
  <td><div align="center"><?php echo $newrow->amoney; ?></div></td>
  <td><div align="center"><?php echo $newrow->bmoney; ?></div></td>
  <td><div align="center"><?php echo $row->a1date; ?></div></td>
  <td><div align="center"><?php echo $row->b1date; ?></div></td>
  <td><div align="center"><?php echo $row->a2date; ?></div></td>
  <td><div align="center"><?php echo $row->b2date; ?></div></td>
  <td><div align="center"><?php echo $row->a3date; ?></div></td>
  <td><div align="center"><?php echo $row->b3date; ?></div></td>
  <td><div align="center"><?php echo $row->a4date; ?></div></td>
  <td><div align="center"><?php echo $row->b4date; ?></div></td>
  <td><div align="center"><?php echo $row->a5date; ?></div></td>
  <td><div align="center"><?php echo $row->b5date; ?></div></td>
  <td><div align="center"><?php echo $row->a6date; ?></div></td>
  <td><div align="center"><?php echo $row->b6date; ?></div></td>
  <td><div align="center"><?php echo $row->a7date; ?></div></td>
  <td><div align="center"><?php echo $row->b7date; ?></div></td>
  <td><div align="center"><?php echo $row->a8date; ?></div></td>
  <td><div align="center"><?php echo $row->b8date; ?></div></td>
  <td><div align="center"><?php echo $row->a9date; ?></div></td>
  <td><div align="center"><?php echo $row->b9date; ?></div></td>
  <td><div align="center"><?php echo $row->a10date; ?></div></td>
```

```php
<td><div align="center"><? php echo $row->b10date;?></div></td>
<td><div align="center"><? php echo $row->a11date;?></div></td>
<td><div align="center"><? php echo $row->b11date;?></div></td>
<td><div align="center"><? php echo $row->a12date;?></div></td>
<td><div align="center"><? php echo $row->b12date;?></div></td>
<td><div align="center"><? php echo $row->amoney;?></div></td>
<td><div align="center"><? php echo $row->bmoney;?></div></td>
<td><div align="center"><? php echo $allrow->allmoney;?></div></td>
<td><div align="center">
    <a href="ybmodify.php?ybid=<? php echo $row->ybid;?>&yeadate=<? php echo $pgroup;?>" target="_blank">修改</a>
    <a href="ybdelete.php?ybid=<? php echo $row->ybid;?>&yeadate=<? php echo $pgroup;?>" target="_blank">删除</a>
</div></td>
</tr>
<? php
$i=$i+1;
if($i==$pagesize)
    break;
}
?>
<tr>
<!--合计内容-->
<td><div align="center"></div></td>
<td><div align="center">合计</div></td>
<td><div align="center"><? php echo $hjold;?></div></td>
<td><div align="center"><? php echo $hjaold;?></div></td>
<td><div align="center"><? php echo $hjbold;?></div></td>
<td><div align="center"><? php echo $hja1;?></div></td>
<td><div align="center"><? php echo $hjb1;?></div></td>
<td><div align="center"><? php echo $hja2;?></div></td>
<td><div align="center"><? php echo $hjb2;?></div></td>
<td><div align="center"><? php echo $hja3;?></div></td>
<td><div align="center"><? php echo $hjb3;?></div></td>
<td><div align="center"><? php echo $hja4;?></div></td>
<td><div align="center"><? php echo $hjb4;?></div></td>
<td><div align="center"><? php echo $hja5;?></div></td>
<td><div align="center"><? php echo $hjb5;?></div></td>
<td><div align="center"><? php echo $hja6;?></div></td>
<td><div align="center"><? php echo $hjb6;?></div></td>
<td><div align="center"><? php echo $hja7;?></div></td>
<td><div align="center"><? php echo $hjb7;?></div></td>
<td><div align="center"><? php echo $hja8;?></div></td>
<td><div align="center"><? php echo $hjb8;?></div></td>
<td><div align="center"><? php echo $hja9;?></div></td>
<td><div align="center"><? php echo $hjb9;?></div></td>
<td><div align="center"><? php echo $hja10;?></div></td>
<td><div align="center"><? php echo $hjb10;?></div></td>
<td><div align="center"><? php echo $hja11;?></div></td>
<td><div align="center"><? php echo $hjb11;?></div></td>
<td><div align="center"><? php echo $hja12;?></div></td>
```

```php
        <td><div align="center"><?php echo $hjb12;?></div></td>
        <td><div align="center"><?php echo $hjamoney;?></div></td>
        <td><div align="center"><?php echo $hjbmoney;?></div></td>
        <td><div align="center"><?php echo $hjallmoney;?></div></td>
        <td><div align="center">
          </div></td>
      </tr>
    </table>
    <?php
      $result->free();
      $link->close();
    ?>
  <div align="center">
  [第<?php echo $pageno;?>页/共<?php echo $pagecount;?>页]
  <?php
    $href=$PHP_SELF."?work=".urlencode(1);
    if($pageno<>1){
    ?>
      <a href="<?php echo $href;?>&pageno=1&pgroup=<?php echo $pgroup;?>">首页</a>
      <a href="<?php echo $href;?>&pageno=<?php echo $pageno-1;?>&pgroup=<?php echo $pgroup;?>">上一页</a>
    <?php
    }
    if($pageno<>$pagecount){
    ?>
      <a href="<?php echo $href;?>&pageno=<?php echo $pageno+1;?>&pgroup=<?php echo $pgroup;?>">下一页</a>
      <a href="<?php echo $href;?>&pageno=<?php echo $pagecount;?>&pgroup=<?php echo $pgroup;?>">尾页</a>
    <?php
    }
    ?>
    [共<?php echo $rows;?>条信息]
  </div>
  </body>
</html>
```

任务二　数据导入 Word

使用 PHP 读取数据库数据,并将数据导入 Word 文档,具体代码如下:

```
<HTML>
<HEAD>
<title>
</title>
</HEAD>
<body>
<form id="form">
<table id = "PrintA" width="100%" border="1" cellspacing="0" cellpadding="0">
```

```html
<TR style="text-align:center;">
<TD>单元格1</TD>
<TD>单元格2</TD>
<TD>单元格3</TD>
<TD>单元格4</TD>
</TR>
<TR>
<TD colSpan=4 style="text-align:center;"><font color="red" face="Verdana">单元格合并</FONT></TD>
</TR>
</TABLE>
<BR>
<table id="Test" width="100%">
<tr>
<td><font color="red">test</FONT></td>
</tr>
</table>
</form>
<input type="button" onclick="javascript:MakeWord();" value="导出页面到Word">
<SCRIPT LANGUAGE="javascript">
function MakeWord()
{
var word = new ActiveXObject("Word.Application");
// var doc = word.documents.open("c:test.doc");    //此处为打开已有的模版
var doc = word.Documents.Add("",0,1);//不打开模版直接加入内容
var Range=doc.Range();
var sel = document.body.createTextRange();
sel.moveToElementText(form);//此处form是页面form的id
sel.select();

sel.execCommand("Copy");
Range.Paste();
word.Application.Visible = true;
alert("s");
word.Application.Selection.InlineShapes.AddPicture("c:\m20.gif");
alert("n");
doc.saveAs("c:\ba.doc");      //存放到指定的位置注意路径一定要是"\"不然会报错
}
</SCRIPT>
</body>
</html>
```

五、考核标准

(1) 版面布局合理清晰,整体效果美观,观赏性强。(10分)

(2) 网页中没有明显的错误(如超链接、图片无法显示、错别字等)。(10分)

(3) 读取数据库数据,以表格形式出现在网页中。(20分)

(4) 将数据导入Excel。(30分)

(5) 将数据导入Word。(30分)

项目十一　jQuery 应用

一、实训目的

- 掌握 jQuery 的基本语法；
- 掌握 jQuery 的动态效果。

二、实训要求

- 下载 jQuery，并在网页中正确引用；
- 使用 jQuery，创建网页动态效果。

三、实训设计

jQuery 是一个 JavaScript 函数库。如需使用 jQuery，您需要下载 jQuery 库，然后把它包含在希望使用的网页中。共有两个版本的 jQuery 可供下载：一份是精简过的；另一份是未压缩的（供调试或阅读）。这两个版本都可从 jQuery.com 下载。

本实训通过引用 jQuery，简化 Javascript 代码的书写量，创建出漂亮的页面动态效果。

四、实训内容

任务一　jQuery 引用

jQuery 库位于一个 JavaScript 文件中，其中包含了所有的 jQuery 函数。可以通过下面的标记把 jQuery 添加到网页中：

```
<head>
<script type="text/javascript" src="jquery.js"></script>
</head>
```

另外，Google 和 Microsoft 对 jQuery 的支持都很好。如果您不愿意在自己的计算机上存放 jQuery 库，那么可以从 Google 或 Microsoft 加载 CDN jQuery 核心文件。

使用 Google 的 CDN：

```
<head>
<script type="text/javascript" src="http://ajax.googleapis.com/ajax/libs/jquery/1.4.0/jquery.min.js"></script>
</head>
```

使用 Microsoft 的 CDN：

```
<head>
<script type="text/javascript" src="http://ajax.microsoft.com/ajax/jquery/jquery-1.4.min.js"></script>
</head>
```

任务二 jQuery 隐藏显示

方法一：使用 jQuery 类库中的 hide()和 show()方法来隐藏和显示 HTML 元素，代码如下：

```
<! DOCTYPE html>
<html>
<head>
<script src="/jquery/jquery-1.11.1.min.js"></script>
<script type="text/javascript">
$(document).ready(function(){
  $("#hide").click(function(){
  $("p").hide();
  });
  $("#show").click(function(){
  $("p").show();
  });
});
</script>
</head>
<body>
<p id="p1">如果点击"隐藏"按钮，我就会消失。</p>
<button id="hide" type="button">隐藏</button>
<button id="show" type="button">显示</button>
</body>
</html>
```

运行效果见图 11.1。

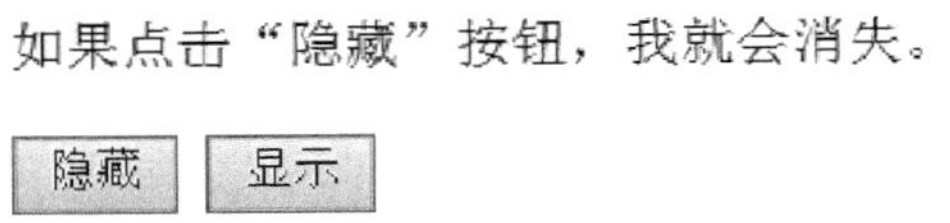

如果点击"隐藏"按钮，我就会消失。

隐藏　显示

图 11.1　隐藏显示

方法二：

在隐藏消失功能中，可以使用 speed 参数控制隐藏/显示的速度，可以取值："slow"、"fast" 或毫秒。要求当点击隐藏按钮后，按钮下面的文字将会慢慢消失，代码如下：

```
<! DOCTYPE html>
<html>
<head>
<script src="/jquery/jquery-1.11.1.min.js"></script>
<script type="text/javascript">
$(document).ready(function(){
  $("button").click(function(){
  $("p").hide(1000);
  });
});
</script>
</head>
```

```
<body>
<button type="button">隐藏</button>
<p>这是一个段落。</p>
<p>这是另一个段落。</p>
</body>
</html>
```

效果见图 11.2。

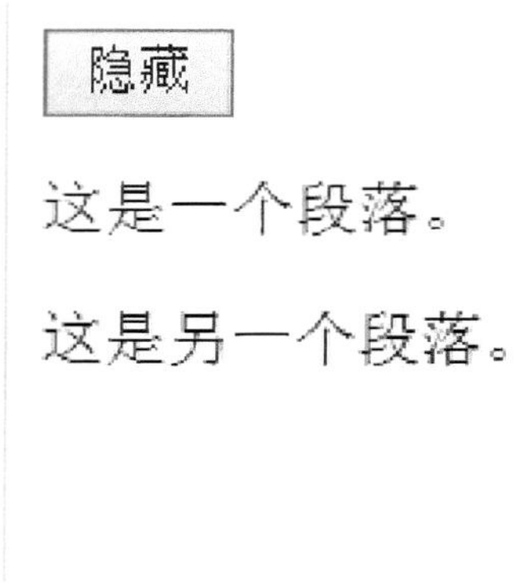

图 11.2　缓慢隐藏

任务三　jQuery 淡入淡出

要求通过 jQuery 实现点击网页上的按钮后,三个矩阵慢慢淡出,实现具体代码如下:
方法一:

```
<html>
<head>
<script src="/jquery/jquery-1.11.1.min.js"></script>
<script>
$(document).ready(function(){
    $("button").click(function(){
        $("#div1").fadeToggle();
        $("#div2").fadeToggle("slow");
        $("#div3").fadeToggle(3000);
    });
});
</script>
</head>
<body>
<p>演示带有不同参数的 fadeToggle() 方法。</p>
<button>点击这里,使三个矩形淡入淡出</button>
<br><br>
<div id="div1" style="width:80px;height:80px;background-color:red;"></div>
<br>
<div id="div2" style="width:80px;height:80px;background-color:green;"></div>
<br>
<div id="div3"
```

```
style="width:80px;height:80px;background-color:blue;"></div>
    </body>
  </body>
</html>
```

代码运行效果见图 11.3。

图 11.3 矩阵淡出

方法二：
下面是可控透明度的淡入淡出：

```
<!DOCTYPE html>
<html>
<head>
<script src="/jquery/jquery-1.11.1.min.js"></script>
<script>
$(document).ready(function(){
  $("button").click(function(){
    $("#div1").fadeTo("slow",0.15);
    $("#div2").fadeTo("slow",0.4);
    $("#div3").fadeTo("slow",0.7);
  });
});
</script>
</head>
<body>
<p>演示带有不同参数的 fadeTo() 方法。</p>
<button>点击这里,使三个矩形淡出</button>
<br><br>
<div id="div1" style="width:80px;height:80px;background-color:red;"></div>
<br>
<div id="div2" style="width:80px;height:80px;background-color:green;"></div>
<br>
<div id="div3" style="width:80px;height:80px;background-color:blue;"></div>
</body>
</html>
```

效果见图 11.4。

图 11.4 颜色淡出

任务四　jQuery 滑动效果

通过 jQuery 的 slideDown()、slideUp()、slideToggle()等方法，在元素上创建滑动效果，代码如下：

```
<!DOCTYPE html>
<html>
<head>
<script src="/jquery/jquery-1.11.1.min.js"></script>
<script type="text/javascript">
$(document).ready(function(){
 $(".flip").click(function(){
     $(".panel").slideToggle("slow");
 });
});
</script>

<style type="text/css">
div.panel,p.flip
{
margin:0px;
padding:5px;
text-align:center;
background:#e5eecc;
border:solid 1px #c3c3c3;
}
div.panel
{
height:120px;
display:none;
```

```
}
</style>
</head>

<body>

<div class="panel">
<p>W3School - 领先的 Web 技术教程站点</p>
<p>在 W3School,你可以找到你所需要的所有网站建设教程。</p>
</div>

<p class="flip">请点击这里</p>

</body>
</html>
```

代码运行效果见图 11.5。

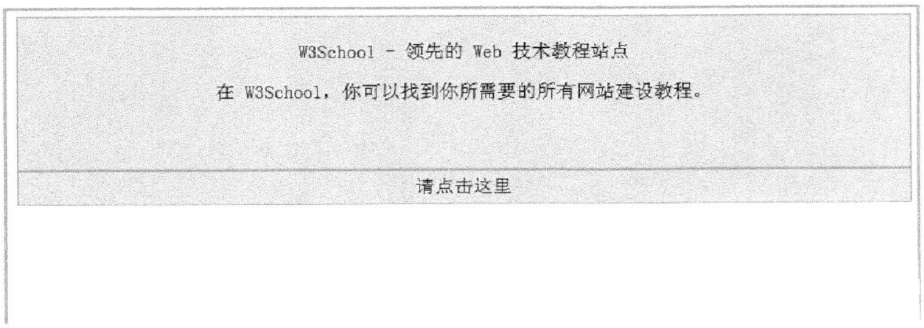

图 11.5 滑动

任务五 jQuery 动画

通过 jQuery 的 animate() 方法,创建自定义动画,代码如下所示:

```
<!DOCTYPE html>
<html>
<head>
<script src="/jquery/jquery-1.11.1.min.js"></script>
<script>
$(document).ready(function(){
  $("button").click(function(){
    var div=$("div");
    div.animate({height:'300px',opacity:'0.4'},"slow");
    div.animate({width:'300px',opacity:'0.8'},"slow");
    div.animate({height:'100px',opacity:'0.4'},"slow");
    div.animate({width:'100px',opacity:'0.8'},"slow");
  });
});
</script>
</head>
<body>
```

```
<button>开始动画</button>
<p>默认情况下,所有 HTML 元素的位置都是静态的,并且无法移动。如需对位置进行操作,记得首先把元素的 CSS position 属性设置为 relative、fixed 或 absolute。</p>
<div style="background:#98bf21;height:100px;width:100px;position:absolute;">
</div>
</body>
</html>
```

代码运行效果见图11.6。

图 11.6　动画

任务六　　jQuery 半透明遮罩效果

jQuery 打造鼠标悬停图片时的半透明遮罩效果,当你的鼠标移上图片的时候,半透明的遮罩层从上到下滑出,覆盖在图片上,并显示出图片的标题和描述。

```
<html xmlns="http://www.w3.org/1999/xhtml">
<head>
<title>jQuery 打造鼠标悬停图片时的半透明遮罩效果</title>
<style>
*{margin:0;padding:0;}
li{list-style:none;}
.box{width:800px;height:400px;margin:50px auto;overflow:hidden;}
.box ul li{width:200px;height:200px;float:left;position:relative;overflow:hidden;}
.box ul li .dask{width:170px;height:180px;padding:20px 0 0 30px;background:#000;opacity:0.8;position:absolute;top:-200px;left:0;}
.box ul li .dask p{color:#fff;}
.box ul li .dask a{color:green;text-decoration:none}
</style>
<script type="text/javascript" src="/ajaxjs/jquery-1.9.1.min.js"></script>
<script>
$(function(){
    $(".box ul li").hover(
        function(){
```

```
                $(this).find(".dask").stop().delay(50).animate({"top":0,opacity:0.8},300)
            },
            function () {
                $(this).find(".dask").stop().animate({"top":-200,opacity:0},300)
                        }
    
        )
    })
</script>
</head>
<body>
<div class="box">
<ul>
    <li>
        <a href="/"><img src="/jscss/demoimg/wall_s1.jpg" width="200" height="200" alt="" /></a>
        <div class="dask">
            <p>荃银高科</p>
            <p>天穿</p>
            <a href="/">您想不到的美景</a>
        </div>
    </li>
    <li>
        <a href="/"><img src="/jscss/demoimg/wall_s2.jpg" width="200" height="200" alt="" /></a>
        <div class="dask">
            <p>落日</p>
            <p>天错时</p>
            <a href="/">自然美</a>
        </div>
    </li>
</ul>
</div>
</body>
</html>
```

代码运行效果见图 11.7。

图 11.7 半透明罩

任务七 jQuery 层的拖动

jQuery 拖动 div、移动 div、弹出层实现原理及示例,实现原理是使 div 的 position 为绝对定位 absolute,然后控制其 top 与 left 值,需要监听鼠标事件,主要用到 mousedown,mousemove, mouseup。在 mousedown 后,记录 mousedown 时鼠标与需要移动的 div 的位置,然后取得两者之差,得到在鼠标移动后,div 的位置。即:left=当前鼠标位置.x—(鼠标点击时的.x 值—div 的初始位置 x 值),top=当前鼠标位置.y—(鼠标点击时的.y 值— div 的初始位置 y 值)。具体代码如下:

```html
<!DOCTYPE html>
<html lang="zh">
<head>
<title>jQuery 拖动 div、移动 div、弹出层示例</title>
<script src="http://www.codefans.net/ajaxjs/jquery-1.9.1.min.js" type="text/javascript">
</script>
<style type="text/css">
.moveBar {
position: absolute;
width: 250px;
height: 300px;
background: #666;
border: solid 1px #000;
}
#banner {
background: #52CCCC;
cursor: move;
}
</style>
</head>
<body style="padding-top: 50px;">
<div class="moveBar">
<div id="banner">按住这里可移动当前 div</div>
<div class="content">弹出层的正文内容</div>
</div>
<script>
jQuery(document).ready(
function () {
$('#banner').mousedown(
function (event) {
var isMove = true;
var abs_x = event.pageX - $('div.moveBar').offset().left;
var abs_y = event.pageY - $('div.moveBar').offset().top;
$(document).mousemove(function (event) {
if (isMove) {
var obj = $('div.moveBar');
obj.css({'left':event.pageX - abs_x, 'top':event.pageY - abs_y});
}
}
```

```
).mouseup(
function () {
isMove = false;
}
);
}
);
}
);
</script>
</body>
</html>
```

代码运行效果见图 11.8。

图 11.8　层的拖动

任务八　jQuery 下拉导航菜单

用 jQuery 实现向下滑出的二级菜单代码，滑出菜单，鼠标放在主菜单的任意一项上，就会向下滑出二级的子菜单。

```
<!DOCTYPE html PUBLIC "-//W3C//DTD XHTML 1.0 Transitional//EN"
"http://www.w3.org/TR/xhtml1/DTD/xhtml1-transitional.dtd">
<html xmlns="http://www.w3.org/1999/xhtml">
<head>
<title>jQuery 缓慢弹出下拉导航</title>
<style>
*{margin:0;padding:0;list-style-type:none;}
a,img{border:0;text-decoration:none;}
body{ font: 12px/180% Arial, Helvetica, sans-serif, "新宋体";
background-color: #E8E8E8; }
```

```css
.clearfix:after{content:".";display:block;height:0;clear:both;visibility:hidden}
.clearfix{display:inline-table}
* html .clearfix{height:1%}
.clearfix{display:block}
*+html .clearfix{min-height:1%}
/* nav_menu */
.nav_menu{ height: 42px; background-color: #333333; }
.nav{width:1006px;height:41px;position:relative;margin:0 auto;}
.nav .list li{float:left;}
.nav .list a{float:left;display:block;width:125px;height:42px;text-align:center;font:bold 13px/36px "微软雅黑";color:#fff;}
.nav .list a:hover{color:#FFA304;}
.nav .list a:hover,.nav .list .now{color:#F00;background:#fff;}
.nav .box{position:absolute;left:-5px;top:42px;width:1006px;background:#FFF;overflow:hidden;height:0;filter:alpha(opacity=0);opacity:0;border-bottom:2px solid #074c52;}
.nav .cont{position:relative;padding:25px 0 0px 24px;}
/* sublist */
.sublist li{float:left;width:168px;padding-right:24px;padding-bottom:24px;}
.sublist li h3.mcate-item-hd{font-family:'微软雅黑';padding-left:2px;font-size:14px;height:26px;line-height:26px;border-bottom:1px dashed #666666;}
.sublist li p.mcate-item-bd{padding-left:2px;}
.sublist li p.mcate-item-bd a{height:26px;line-height:26px;margin-right:5px;font-size:12px;color:#666666;text-decoration:none;display:inline-block;}
.sublist li p.mcate-item-bd a:hover{color:#6c5143;text-decoration:underline;}
</style>
<script type="text/javascript" src="/ajaxjs/jquery-1.6.2.min.js"></script>
</head>
<body>
<div class="nav_menu">
    <div class="nav">
        <div class="list" id="navlist">
            <ul id="navfouce">
                <li><a href="/">公司概况</a></li>
                <li><a href="/">产品展示</a></li>
                <li><a href="/">新闻动态</a></li>
                <li><a href="/">营销网络</a></li>
                <li><a href="/">照明知识</a></li>
                <li><a href="/">人力资源</a></li>
                <li><a href="/">客服中心</a></li>
                <li><a href="/">联系我们</a></li>
            </ul>
        </div>
        <div class="box" id="navbox" style="height:0px;opacity:0;overflow:hidden;">
            <div class="cont" style="display:none;">
                <ul class="sublist clearfix">
                    <li>
```

```html
                <h3 class="mcate-item-hd"><span>服饰内衣</span></h3>
                <p class="mcate-item-bd">
                    <a href="/">女装</a>
                    <a href="/">男装</a>
                    <a href="/">内衣</a>
                    <a href="/">家居服</a>
                    <a href="/">配件</a>
                    <a href="/">羽绒</a>
                    <a href="/">呢大衣</a>
                    <a href="/">毛衣</a>
                </p>
            </li>
            <li>
                <h3 class="mcate-item-hd"><span>鞋 箱包</span></h3>
                <p class="mcate-item-bd">
                    <a href="/">女鞋</a>
                    <a href="/">男鞋</a>
                    <a href="/">箱包</a>
                    <a href="/">女包</a>
                    <a href="/">男包</a>
                    <a href="/">旅行箱</a>
                    <a href="/">钱包 </a>
                </p>
            </li>
            <li>
                <h3 class="mcate-item-hd"><span>珠宝、手表</span></h3>
                <p class="mcate-item-bd">
                    <a href="/">饰品</a>
                    <a href="/">项链</a>
                    <a href="/">珠宝</a>
                    <a href="/">钻石</a>
                    <a href="/">手表</a>
                </p>
            </li>
            <li>
                <h3 class="mcate-item-hd"><span>化妆品</span></h3>
                <p class="mcate-item-bd">
                    <a href="/">护肤</a>
                    <a href="/">彩妆</a>
                    <a href="/">香水</a>
                    <a href="/">男士</a>
                    <a href="/">精油</a>
                    <a href="/">假发</a>
                    <a href="/">美体</a>
                    <a href="/">试用服务</a>
                </p>
            </li>
            <li>
                <h3 class="mcate-item-hd"><span>运动 户外</span></h3>
                <p class="mcate-item-bd">
```

```html
            <a href="/">运动鞋</a>
            <a href="/">运动服</a>
            <a href="/">运动用品</a>
            <a href="/">户外</a>
        </p>
    </li>
    <li>
        <h3 class="mcate-item-hd"><span>手机 数码</span></h3>
        <p class="mcate-item-bd">
            <a href="/">手机</a>
            <a href="/">笔记本</a>
            <a href="/">相机</a>
            <a href="/">平板电脑</a>
            <a href="/">配件</a>
            <a href="/">电脑硬件</a>
        </p>
    </li>
    <li>
        <h3 class="mcate-item-hd"><span>家用电器</span></h3>
        <p class="mcate-item-bd">
            <a href="/">大家电</a>
            <a href="/">影音电器</a>
            <a href="/">生活电器</a>
            <a href="/">厨房电器</a>
            <a href="/">健康护理</a>
            <a href="/">剃须刀</a>
        </p>
    </li>
    <li>
        <h3 class="mcate-item-hd"><span>家具 建材</span></h3>
        <p class="mcate-item-bd">
            <a href="/">家具</a>
            <a href="/">卫浴</a>
            <a href="/">地板</a>
            <a href="/">灯具</a>
            <a href="/">五金</a>
            <a href="/">开关</a>
            <a href="/">装修设计</a>
        </p>
    </li>
    <li>
        <h3 class="mcate-item-hd"><span>家纺 居家</span></h3>
        <p class="mcate-item-bd">
            <a href="/">家纺</a>
            <a href="/">磨毛套件</a>
            <a href="/">羽绒被</a>
            <a href="/">枕头</a>
            <a href="/">软饰</a>
            <a href="/">居家</a>
            <a href="/">厨房</a>
        </p>
```

```html
        </li>
        <li>
<h3 class="mcate-item-hd"><span>食品</span></h3>
    <p class="mcate-item-bd">
        <a href="/">零食</a>
        <a href="/">进口</a>
        <a href="/">茶叶</a>
        <a href="/">冲饮</a>
        <a href="/">酒水</a>
        <a href="/">粮油</a>
        <a href="/">干货</a>
        <a href="/">生鲜</a>
    </p>
</li>
<li>
<h3 class="mcate-item-hd"><span>医药保健</span></h3>
    <p class="mcate-item-bd">
        <a href="/">保健</a>
        <a href="/">滋补</a>
        <a href="/">蛋白粉</a>
        <a href="/">阿胶</a>
        <a href="/">药品</a>
        <a href="/">血压仪</a>
        <a href="/">计生</a>
        <a href="/">体检</a>
    </p>
</li>
<li>
<h3 class="mcate-item-hd"><span>母婴用品</span></h3>
    <p class="mcate-item-bd">
    <a href="/">玩具</a>
        <a href="/">宝宝食品</a>
        <a href="/">用品</a>
        <a href="/">童装</a>
        <a href="/">孕装</a>
    </p>
</li>
<li>
<h3 class="mcate-item-hd"><span>汽车配件</span></h3>
    <p class="mcate-item-bd">
        <a href="/">新车</a>
        <a href="/">座垫</a>
        <a href="/">脚垫</a>
    <a href="/">GPS</a>
    <a href="/">车衣</a>
        <a href="/">洗车机</a>
        <a href="/">水枪</a>
    </p>
</li>
<li>
<h3 class="mcate-item-hd"><span>文化玩乐</span></h3>
```

```html
                <p class="mcate-item-bd">
                    <a href="/">电子凭证</a>
                    <a href="/">图书</a>
                    <a href="/">乐器</a>
                    <a href="/">旅游</a>
                    <a href="/">鲜花</a>
                </p>
            </li>
        </ul>
    </div>
    <div class="cont" style="display:none;">
        <ul class="sublist clearfix">
            <li>
                <h3 class="mcate-item-hd"><span>服饰内衣</span></h3>
                <p class="mcate-item-bd">
                    <a href="/">女装</a>
                    <a href="/">男装</a>
                    <a href="/">内衣</a>
                    <a href="/">家居服</a>
                    <a href="/">配件</a>
                    <a href="/">羽绒</a>
                    <a href="/">呢大衣</a>
                    <a href="/">毛衣</a>
                </p>
            </li>
            <li>
                <h3 class="mcate-item-hd"><span>鞋 箱包</span></h3>
                <p class="mcate-item-bd">
                    <a href="/">女鞋</a>
                    <a href="/">男鞋</a>
                    <a href="/">箱包</a>
                    <a href="/">女包</a>
                    <a href="/">男包</a>
                    <a href="/">旅行箱</a>
                    <a href="/">钱包 </a>
                </p>
            </li>
            <li>
                <h3 class="mcate-item-hd"><span>珠宝、手表</span></h3>
                <p class="mcate-item-bd">
                    <a href="/">饰品</a>
                    <a href="/">项链</a>
                    <a href="/">珠宝</a>
                    <a href="/">钻石</a>
                    <a href="/">手表</a>
                </p>
            </li>
            <li>
                <h3 class="mcate-item-hd"><span>化妆品</span></h3>
                <p class="mcate-item-bd">
                    <a href="/">护肤</a>
```

```html
            <a href="/">彩妆</a>
            <a href="/">香水</a>
            <a href="/">男士</a>
            <a href="/">精油</a>
            <a href="/">假发</a>
            <a href="/">美体</a>
            <a href="/">试用服务</a>
        </p>
    </li>
    <li>
        <h3 class="mcate-item-hd"><span>运动 户外</span></h3>
        <p class="mcate-item-bd">
            <a href="/">运动鞋</a>
            <a href="/">运动服</a>
            <a href="/">运动用品</a>
            <a href="/">户外</a>
        </p>
    </li>
    <li>
        <h3 class="mcate-item-hd"><span>手机 数码</span></h3>
        <p class="mcate-item-bd">
            <a href="/">手机</a>
            <a href="/">笔记本</a>
            <a href="/">相机</a>
            <a href="/">平板电脑</a>
            <a href="/">配件</a>
            <a href="/">电脑硬件</a>
        </p>
    </li>
    <li>
        <h3 class="mcate-item-hd"><span>家用电器</span></h3>
        <p class="mcate-item-bd">
            <a href="/">大家电</a>
            <a href="/">影音电器</a>
            <a href="/">生活电器</a>
            <a href="/">厨房电器</a>
            <a href="/">健康护理</a>
            <a href="/">剃须刀</a>
        </p>
    </li>
    <li>
        <h3 class="mcate-item-hd"><span>家具 建材</span></h3>
        <p class="mcate-item-bd">
            <a href="/">家具</a>
            <a href="/">卫浴</a>
            <a href="/">地板</a>
            <a href="/">灯具</a>
            <a href="/">五金</a>
            <a href="/">开关</a>
            <a href="/">装修设计</a>
        </p>
```

```html
        </li>
        <li>
<h3 class="mcate-item-hd"><span>家纺 居家</span></h3>
    <p class="mcate-item-bd">
        <a href="/">家纺</a>
        <a href="/">磨毛套件</a>
        <a href="/">羽绒被</a>
        <a href="/">枕头</a>
        <a href="/">软饰</a>
        <a href="/">居家</a>
        <a href="/">厨房</a>
    </p>
        </li>
        <li>
<h3 class="mcate-item-hd"><span>食品</span></h3>
        <p class="mcate-item-bd">
            <a href="/">零食</a>
            <a href="/">进口</a>
            <a href="/">茶叶</a>
            <a href="/">冲饮</a>
            <a href="/">酒水</a>
            <a href="/">粮油</a>
            <a href="/">干货</a>
            <a href="/">生鲜</a>
        </p>
        </li>
    </ul>
</div>
<div class="cont" style="display:none;">
    <ul class="sublist clearfix">
        <li>
<h3 class="mcate-item-hd"><span>服饰内衣</span></h3>
    <p class="mcate-item-bd">
        <a href="/">女装</a>
        <a href="/">男装</a>
        <a href="/">内衣</a>
        <a href="/">家居服</a>
        <a href="/">配件</a>
        <a href="/">羽绒</a>
        <a href="/">呢大衣</a>
        <a href="/">毛衣</a>
        </p>
        </li>
        <li>
<h3 class="mcate-item-hd"><span>鞋 箱包</span></h3>
    <p class="mcate-item-bd">
        <a href="/">女鞋</a>
        <a href="/">男鞋</a>
        <a href="/">箱包</a>
        <a href="/">女包</a>
        <a href="/">男包</a>
```

```html
                    <a href="/">旅行箱</a>
                    <a href="/">钱包 </a>
                </p>
            </li>
            <li>
                <h3 class="mcate-item-hd"><span>珠宝、手表</span></h3>
                <p class="mcate-item-bd">
                    <a href="/">饰品</a>
                    <a href="/">项链</a>
                    <a href="/">珠宝</a>
                    <a href="/">钻石</a>
                    <a href="/">手表</a>
                </p>
            </li>
            <li>
                <h3 class="mcate-item-hd"><span>化妆品</span></h3>
                <p class="mcate-item-bd">
                    <a href="/">护肤</a>
                    <a href="/">彩妆</a>
                    <a href="/">香水</a>
                    <a href="/">男士</a>
                    <a href="/">精油</a>
                    <a href="/">假发</a>
                    <a href="/">美体</a>
                    <a href="/">试用服务</a>
                </p>
            </li>
            <li>
                <h3 class="mcate-item-hd"><span>运动 户外</span></h3>
                <p class="mcate-item-bd">
                    <a href="/">运动鞋</a>
                    <a href="/">运动服</a>
                    <a href="/">运动用品</a>
                    <a href="/">户外</a>
                </p>
            </li>
        </ul>
    </div>
    <div class="cont" style="display:none;">3<br />3</div>
    <div class="cont" style="display:none;">4<br />3<br />4</div>
    <div class="cont" style="display:none;">5</div>
    <div class="cont" style="display:none;">6<br />3<br />3</div>
    <div class="cont" style="display:none;">7<br />3<br />3<br />3</div>
    </div>
</div>
</div>
<script type="text/javascript">
(function(){
    var time = null;
    var list = $("#navlist");
    var box = $("#navbox");
```

```javascript
        var lista = list.find("a");
        for(var i=0,j=lista.length;i<j;i++){
            if(lista[i].className == "now"){
                var olda = i;
            }
        }
        var box_show = function(hei){
            box.stop().animate({
                height:hei,
                opacity:1
            },400);
        }
        var box_hide = function(){
            box.stop().animate({
                height:0,
                opacity:0
            },400);
        }
        lista.hover(function(){
            lista.removeClass("now");
            $(this).addClass("now");
            clearTimeout(time);
            var index = list.find("a").index($(this));
            box.find(".cont").hide().eq(index).show();
            var _height = box.find(".cont").eq(index).height()+54;
            box_show(_height)
        },function(){
            time = setTimeout(function(){
                box.find(".cont").hide();
                box_hide();
            },50);
            lista.removeClass("now");
            lista.eq(olda).addClass("now");
        });
        box.find(".cont").hover(function(){
            var _index = box.find(".cont").index($(this));
            lista.removeClass("now");
            lista.eq(_index).addClass("now");
            clearTimeout(time);
            $(this).show();
            var _height = $(this).height()+54;
            box_show(_height);
        },function(){
            time = setTimeout(function(){
                $(this).hide();
                box_hide();
            },50);
            lista.removeClass("now");
            lista.eq(olda).addClass("now");
        });
    })();
```

```
</script>
</body>
</html>
```

代码运行效果见图 11.9。

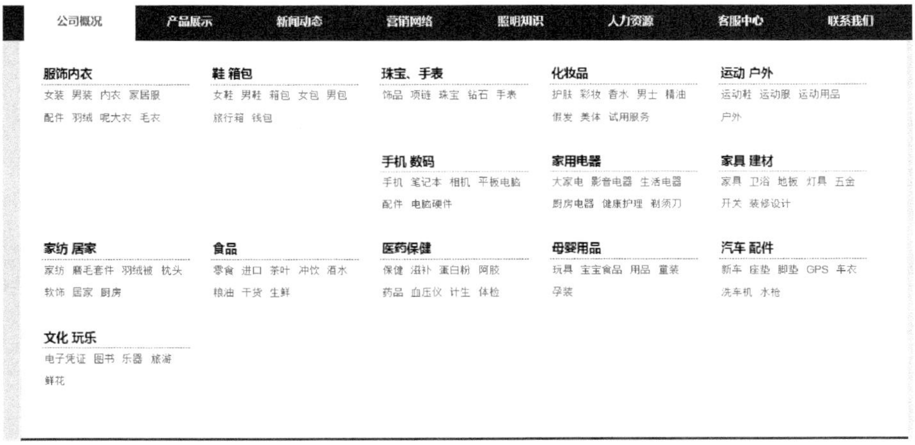

图 11.9 下拉菜单

五、考核标准

（1）版面布局合理清晰，整体效果美观，观赏性强。（10 分）
（2）网页中没有明显的错误（如超链接、图片无法显示、错别字等）。（10 分）
（3）jQuery 引用。（10 分）
（4）jQuery 动态效果。（70 分）

项目十二 教务管理系统

一、实训目的

- 掌握 PHP 系统程序开发的流程；
- 掌握数据库设计的基本用法；
- 掌握 PHP、Html、JavaScript、Apache、MySQL 的综合使用。

二、实训要求

- 设计学生成绩管理系统主页面、登录页面、成绩管理、信息查询等页面；
- 设计学生成绩系统数据库；
- 使用 PHP 开发学生成绩管理系统。

三、实训设计

教务管理系统是学校常用的管理系统，师生在校工作学习过程中，经常使用教务管理系统进行信息查询。同时，教务管理系统也大大简化了学校对师生信息、考务信息的管理和维护，有助于提高学校的管理效率。

➢ 模块设计

（1）用户登录模块：填写已分配的用户名称，填写正确的密码，进入主控制页面。
（2）显示模块：显示要求的内容。
（3）查询模块：提供多种查询条件，可按需要进行查询。
（4）录入模块：向数据库中添加记录。
（5）修改模块：可以找到指定信息并对其进行修改。
（6）删除模块：找到要删除的记录，并将其删除。

四、实训内容

任务一 功能结构图

（1）登录功能见图 12.1。
（2）学生功能见图 12.2。

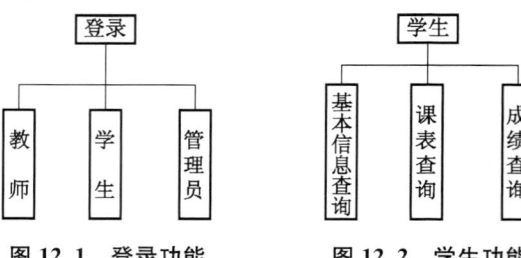

图 12.1　登录功能　　　　图 12.2　学生功能

（3）教师功能见图 12.3。
（4）管理员功能见图 12.4。

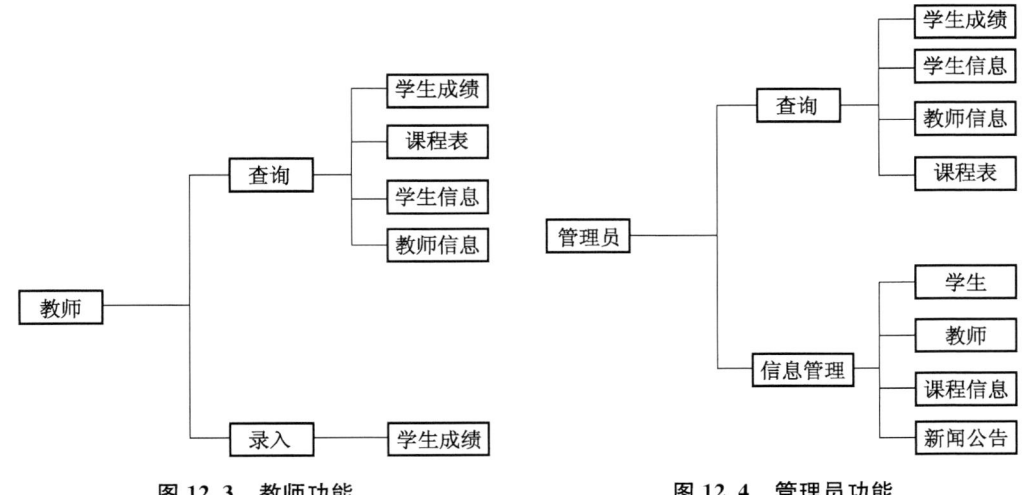

图 12.3　教师功能　　　　　　　图 12.4　管理员功能

任务二　数据库设计

（1）班级表，见图 12.5。

名	类型	长度	十进位	允许空值	
班级代码	int	8	0	☐	🔑1
班级名称	char	6	0	☐	
班级人数	int	10	0	☐	
所属院	varchar	50	0	☑	
所属系	varchar	50	0	☑	

图 12.5　班级表

（2）学生课程表，见图 12.6。

名	类型	长度	十进位	允许空值	
课程名	char	10	0	☐	
班级代码	char	8	0	☐	🔑1
星期	char	60	0	☐	🔑2
节次	char	25	0	☐	🔑3
上课地点	char	12	0	☐	
教工编号	varchar	50	0	☐	
姓名	char	12	0	☑	

图 12.6　学生表

（3）教师课程表，见图 12.7。

名	类型	长度	十进位	允许空值	
班级名称	char	50	0	☑	
节次	char	50	0	☐	🔑1
星期	char	50	0	☐	🔑2
课程名	char	50	0	☑	
教工编号	varchar	50	0	☐	🔑3
上课地点	char	12	0	☐	
学年	varchar	20	0	☑	
学期	varchar	20	0	☑	

图 12.7　教师课程表

(4) 教师职工表，见图 12.8。

名	类型	长度	十进位	允许空值	
教工编号	int	4	0	☐	🔑1
姓名	varchar	50	0	☑	
密码	varchar	20	0	☑	
电话	varchar	16	0	☑	
QQ邮箱	varchar	30	0	☑	
系名	varchar	30	0	☑	
工龄	char	12	0	☑	
身份证号	varchar	50	0	☑	
添加时间	datetime	0	0	☑	

图 12.8　教师职工表

(5) 学生表，见图 12.9。

名	类型	长度	十进位	允许空值	
学号	varchar	20	0	☐	🔑1
密码	varchar	20	0	☑	
姓名	varchar	20	0	☑	
班级代码	varchar	28	0	☑	
性别	char	8	0	☑	
籍贯	varchar	20	0	☑	
电话	varchar	12	0	☑	
QQ邮箱	varchar	20	0	☑	
备注	varchar	50	0	☑	
出生日期	datetime	0	0	☑	
照片	blob	0	0	☑	

图 12.9　学生表

(6) 学生成绩表，见图 12.10。

名	类型	长度	十进位	允许空值	
学号	varchar	20	0	☐	🔑1
姓名	varchar	28	0	☑	
班级代码	varchar	50	0	☑	
学年	varchar	60	0	☑	
学期	char	50	0	☑	
课程名	varchar	50	0	☐	🔑2
成绩	varchar	28	0	☑	
添加时间	datetime	0	0	☑	

图 12.10　学生成绩表

(7) 学生考务表，见图 12.11。

名	类型	长度	十进位	允许空值	
学号	char	12	0	☐	
姓名	varchar	50	0	☐	
班级代码	varchar	28	0	☑	
课程名	char	12	0	☐	🔑1
学年	varchar	50	0	☑	
学期	varchar	50	0	☑	
考试时间	char	12	0	☐	
考试地点	char	16	0	☐	
座位号	int	11	0	☐	

图 12.11　学生考务表

(8) 学生选修表，见图 12.12。
(9) 学生重修表，见图 12.13。

名	类型	长度	十进位	允许空值
学号	char	12	0	
姓名	char	16	0	✓
课程名	char	8	0	
课程号	char	10	0	🔑1
上课地点	char	8	0	✓
上课时间	varchar	6	0	
任课教师	varchar	20	0	✓
学年	varchar	50	0	✓
学期	varchar	50	0	✓

图 12.12　学生选修表

名	类型	长度	十进位	允许空值
学号	char	12	0	🔑1
班级代码	varchar	30	0	✓
班级名称	varchar	50	0	✓
学年	char	10	0	
学期	char	8	0	
课程名	char	12	0	
重修成绩	float	8	0	

图 12.13　学生重修表

（10）用户表，见图 12.14。

名	类型	长度	十进位	允许空值
用户编号	int	6	0	🔑1
用户名	varchar	50	0	✓
密码	varchar	50	0	✓
权限	varchar	50	0	✓
添加时间	datetime	0	0	✓

图 12.14　用户表

（11）专业选修课表，见图 12.15。

名	类型	长度	十进位	允许空值
学年	char	12	0	✓
学期	char	6	0	
课程代码	char	8	0	🔑1
课程名称	char	10	0	
专业类型	char	6	0	✓
任课教师	char	8	0	✓

图 12.15　专业选修课表

（12）院系表，见图 12.16。

名	类型	长度	十进位	允许空值
系别编号	varchar	50	0	🔑1
系名	varchar	50	0	✓
添加时间	date	0	0	✓

图 12.16　院系表

任务三　登录页面

只有登录后，才可进入网站，登录页面，见图 12.17。index.htm 页面代码：

```html
<!DOCTYPE html PUBLIC "-//W3C//DTD XHTML 1.0 Transitional//EN"
"http://www.w3.org/TR/xhtml1/DTD/xhtml1-transitional.dtd">
<html xmlns="http://www.w3.org/1999/xhtml">
<head>
<title>1.jpg</title>
<link rel="stylesheet" type="text/css" href="css\css1.css">
<meta http-equiv="Content-Type" content="text/html;charset=gb2312">
<meta name="description" content="FW MX CSS Layer">
</head>
<body bgcolor="#ffffff">
<div id="L1" >
    <div id="L2r1c1" ><p class="loginTitle" align="center"><br /><font size="10"><br /><br />    河南财政税务高等专科学校   </font></p></div>
    <div id="L2r2c1" >
        <div id="L2r2c1r1c1" ><img name="N2_r2_c1_r1_c1" src="images1/2_r2_c1_r1_c1.jpg" width="258" height="244" border="0"></div>
        <div id="L2r2c1r1c2" >
            <div id="L2r2c1r1c2r1c1" ><p class="loginTitle" id="loginTitle" ><font size="6">教务管理系统</font></p> </div>
            <div id="L2r2c1r1c2r2c1" >
                <div id="L2r2c1r1c2r2c1r1c1" ><img name="N2_r2_c1_r1_c2_r2_c1_r1_c1" src="images1/2_r2_c1_r1_c2_r2_c1_r1_c1.jpg" width="216" height="179" border="0"></div>
                <div id="L2r2c1r1c2r2c1r1c2" >
                    <div id="L2r2c1r1c2r2c1r1c2r1c1" ><img name="N2_r2_c1_r1_c2_r2_c1_r1_c2_r1_c1" src="images1/2_r2_c1_r1_c2_r2_c1_r1_c2_r1_c1.jpg" width="248" height="27" border="0"></div>
                    <div id="L2r2c1r1c2r2c1r1c2r2c1" >
<form method="get" action="ex10_30.php">
用户名：
<input type="text" name="username" id="username" size="10" /><br /><br />
口  令：
<input type="password" name="pswd" id="pswd" SIZE="10" /><br />
<input name="类型" type="radio" id="js" value="教师" />教师
<input name="类型" type="radio" id="xs" value="学生" checked />学生
<input name="类型" type="radio" id="gly" value="管理员" />管理员
<br /><br />
<input type="submit" value="登录" />
<input type="submit" value="重置" />
</form>
</div></div></div></div>
        <div id="L2r2c1r1c3" ><img name="N2_r2_c1_r1_c3" src="images1/2_r2_c1_r1_c3.jpg" width="233" height="244" border="0"></div></div>
    <div id="L2r3c1" ><img name="N2_r3_c1" src="images1/2_r3_c1.jpg" width="955" height="155" border="0"></div></div>
</body>
</html>
```

ex10_30.php 页面代码：

```php
<? php
session_start();
if($_GET["类型"]=="学生"){
    $username=$_GET["username"];
    $pswd=$_GET["pswd"];
    //连接MySQL服务器,打开数据库
    mysql_connect("127.0.0.1","root","123456");
    mysql_select_db("test");
    //在users表中查找用户
    $query = "SELECT 用户编号,用户名,权限 FROM 用户 WHERE 用户编号=$username AND 密码=$pswd ";
    mysql_query("SET NAMES 'gbk'");
    $result = mysql_query($query);

    //如果找到用户,设置会话变量
    $zhj=mysql_num_rows($result);
    if ($zhj == 1) {
        if(mysql_result($result,0,"权限")==$_GET["类型"]){
            $_SESSION['name'] = mysql_result($result,0,"用户名");
            $_SESSION['username'] = mysql_result($result,0,"用户编号");
            $_SESSION['user'] = mysql_result($result,0,"权限");
            header("refresh:1;url=zhuyemian1.php");
                                    //Location:zhuyemian1.php
        }else{
            echo "身份有误!"."<br>";
            exit;
        }

        exit;
    }
    // 如果未提交表单,显示登录表单
    else {
        echo "用户名或密码输入有误!"."<br>";
        exit;
    }
}
else if($_GET["类型"]=="教师"){
    $username=$_GET["username"];
    $pswd=$_GET["pswd"];
    //连接MySQL服务器,打开数据库
    mysql_connect("127.0.0.1","root","123456");
    mysql_select_db("test");
    //在users表中查找用户
    $query = "SELECT 用户编号,用户名,权限 FROM 用户 WHERE 用户编号=$username AND 密码=$pswd";
    mysql_query("SET NAMES 'gbk'");
    $result = mysql_query($query);
    //如果找到用户,设置会话变量
```

```php
            $zhj=mysql_num_rows($result);
            if ($zhj == 1) {
            if(mysql_result($result,0,"权限") == $_GET["类型"]){
                            $_SESSION['name'] = mysql_result($result,0,"用户名");
                            $_SESSION['username'] = mysql_result($result,0,"用户编号");
                            $_SESSION['user'] = mysql_result($result,0,"权限");
            header("refresh:1;url=zhuyemian2.php");
            //header("Location:zhuyemian2.php");
                    }else{
                        echo "身份有误!"."<br>";
                                    exit;
                    }
                    exit;
            }
            // 如果未提交表单,显示登录表单
        else {
            echo "用户名或密码输入有误!"."<br>";
            exit;
        }
        }
            else if($_GET["类型"]=="管理员"){
                $username = $_GET["username"];
        $pswd = $_GET["pswd"];
        //连接MySQL服务器,打开数据库
        mysql_connect("127.0.0.1","root","123456");
        mysql_select_db("test");
        //在users表中查找用户
            $query = "SELECT 用户编号,用户名,权限 FROM 用户 WHERE 用户编号=$username AND 密码=$pswd";
        mysql_query("SET NAMES 'gbk'");
        $result = mysql_query($query);
        //如果找到用户,设置会话变量
        $zhj=mysql_num_rows($result);
        if ($zhj == 1) {
                            if(mysql_result($result,0,"权限") == $_GET["类型"]){
                        $_SESSION['name'] = mysql_result($result,0,"用户名");
                            $_SESSION['username'] = mysql_result($result,0,"用户编号");
                            $_SESSION['user'] = mysql_result($result,0,"权限");
            header("refresh:1;url=zhuyemian3.php");
            //header("Location:zhuyemian3.php");
                            }else{
                        echo "身份有误!"."<br>";
                                    exit;
                }
                exit;
```

```
        }
        // 如果未提交表单,显示登录表单
    else{
        echo "用户名或密码输入有误!"."<br>";
        exit;
        }
    }
?>
```

代码运行效果见图12.17。

图12.17 登录页面

任务四 学 生 主 页

此页面为学生登录后页面,包含学生能够使用的各种功能导航。
zhuyemian1.php页面代码:

```
<! DOCTYPE html PUBLIC "-//W3C//DTD XHTML 1.0 Transitional//EN"
"http://www.w3.org/TR/xhtml1/DTD/xhtml1-transitional.dtd">
<html xmlns="http://www.w3.org/1999/xhtml">

<head>

<title>1.jpg</title>
<link rel="stylesheet" type="text/css" href="css\css.css">
<meta http-equiv="Content-Type" content="text/html;charset=gb2312">
<meta name="description" content="FW MX CSS Layer">
<style>
a:hover{background:url(images/an.jpg)};
a:visited{ text-decoration:none;};
</style>
</head>
        <? php
            session_start();
            if($_SESSION['user']!="学生"){
            header("Location:index.htm");
            }
```

```html
            ?>
<body bgcolor="#ffffff">
<div id="L1">
    <div id="L2r1c1">

    <p class="loginTitle">
    <font size="8" face="方正桃体">
    <div align="right">
      <font size="3" face="方正桃体" color="#FF0000">你好:
<?php

echo 
    $_SESSION['name'];
?>
  <a href="logout.php">退出</a>
</font>
</div>河南财专教务管理系统</font>


</p>



</div>
<div id="L2r2c1">
        <div id="L2r2c1r1c1">
        <br /><a href="xuexiaoxinxi.html" target="L2r2c1r1c22r1c1">学校基本信息</a>
        <br /><a href="xueshengxinxi.html" target="L2r2c1r1c22r1c1">学生基本信息</a>
        <br /><a href="jiaoshixinxi.html" target="L2r2c1r1c22r1c1">网上报名</a>
        <br /><a href="wangshangpingjia.html" target="L2r2c1r1c22r2c1">网上评价</a>
        <br /><a href="wangshangpingjia.html" target="L2r2c1r1c22r2c1">公告信息</a>
        <br /><a href="http://baidu.com">帮助文件</a></div>
        <div id="L2r2c1r1c2">
            <div id="L2r2c1r1c22r1c1"><iframe name="L2r2c1r1c22r1c1" width="860" height="600" hspace="10" align="right" src="xueshengxinxi.html">您的浏览器不支持浮动框架</iframe></div>
            <div id="L2r2c1r1c22r2c1"><iframe name="L2r2c1r1c22r2c1" width="860" height="600" hspace="10" align="right" src="beijing.html">您的浏览器不支持浮动框架</iframe></div>
</div>
        <div id="L2r3c1" align="center"><img name="N2_r3_c1" src="images/2_r3_c1.jpg" width="1024" height="27" border="0"></div></div>
</body>
</html>
```

学生功能导航页面,xueshengxinxi.html 代码:

```html
<!DOCTYPE html PUBLIC "-//W3C//DTD XHTML 1.0 Transitional//EN"
"http://www.w3.org/TR/xhtml1/DTD/xhtml1-transitional.dtd">
<html xmlns="http://www.w3.org/1999/xhtml">
<head>
<meta http-equiv="Content-Type" content="text/html; charset=gb2312" />
```

```
<title>无标题文档</title>
<style>
a:hover{background:url(images/an.jpg)};
</style>
</head>

<body bgcolor="#66FFFF"><center>
  <a href="xscjcx.php#id" target="L2r2c1r1c22r2c1">学生成绩查询</a>  
      <a href="student_select2_form.php#id" target="L2r2c1r1c22r2c1">学生信息查询</a>  
          <a href="xskb.php#id" target="L2r2c1r1c22r2c1">学生课表查询</a>  
          <a href="xskwcx1.php#id" target="L2r2c1r1c22r2c1">学生考务查询</a>  
          <a href="xsxxkcx.php#id" target="L2r2c1r1c22r2c1">学生选修课情况</a>  
          <a href="xscxcx.php#id" target="L2r2c1r1c22r2c1">学生重修查询</a>
</center>
</body>
</html>
```

代码运行效果见图 12.18。

图 12.18　学生主页

任务五　学生信息查询页面

➢ 学生基本信息查询

查询页面,student_select2_form.php 页面代码:

```
<html>
<head>
<meta http-equiv="Content-Type" content="text/html; charset=gb2312">
<title>学生查询结果</title>
</head>
<body>
<? php
```

```php
$bjdm=$_GET["bjdm"];
$link=new mysqli("127.0.0.1","root","123456");
if(mysqli_connect_errno()) {
    echo "数据库服务器连接失败！<BR>";
    die();
}
$link->select_db("test") or die("数据库选择失败！<BR>");
$link->query("set names 'gbk'");
$sql="select 学号,姓名,性别,班级名称,出生日期 from 学生,班级";
$sql.=" where 学生.班级代码=班级.班级代码 ";
$sql.=" and 学生.班级代码='$bjdm' order by 学号";
$result=$link->query($sql);
$rows=$result->num_rows;  //总记录数
if($rows==0) {
    echo "没有满足条件的记录!";
    die();
}
$pagesize=5;  //每页的记录数(在此暂设为5,通常应设为10)
$pagecount=ceil($rows/$pagesize);  //总页数
//$pageno 的值为当前页的页号
if(!isset($pageno)||$pageno<1)
    $pageno=1;
if($pageno>$pagecount)
    $pageno=$pagecount;
$offset=($pageno-1)*$pagesize;
$result->data_seek($offset);
?>
<div align="center"><strong>学生查询结果</strong> </div>
<table width="90%" border="1" align="center">
  <tr>
    <td><div align="center">学号</div></td>
    <td><div align="center">姓名</div></td>
    <td><div align="center">性别</div></td>
    <td><div align="center">班级</div></td>
    <td><div align="center">操作</div></td>
  </tr>
<?php
$i=0;
while($row=$result->fetch_object()) {
?>
  <tr>
    <td><div align="center"><?php echo $row->学号;?></div></td>
    <td><div align="center"><?php echo $row->姓名;?></div></td>
    <td><div align="center"><?php echo $row->性别;?></div></td>
    <td><div align="center"><?php echo $row->班级名称;?></div></td>
    <td><div align="center">
        <a href="student_detail2.php?xh=<?php echo $row->学号;?>" target="L2r2c1r1c22r2c1">详情</a>
        <a href="student_update_edit.php?xh=<?php echo $row->学号;?>" target="L2r2c1r1c22r2c1">修改</a>
        <a href="student_delete1.php?xh=<?php echo $row->学号;?>" target=
```

```php
"L2r2c1r1c22r2c1">删除</a>
                <a href="student_insertzp.php?xh=<?php echo $row->学号;?>" target="L2r2c1r1c22r2c1">插入</a>
            </div></td>
        </tr>
<?php
    $i=$i+1;
    if($i==$pagesize)
        break;
}
$result->free();
$link->close();
?>
</table>
<div align="center">
[第<?php echo $pageno;?>页/共<?php echo $pagecount;?>页]
<?php
$href=$PHP_SELF."?bjdm=".urlencode($bjdm);
if($pageno<>1){
?>
    <a href="<?php echo $href;?>&pageno=1">首页</a>
    <a href="<?php echo $href;?>&pageno=<?php echo $pageno-1;?>">上一页</a>
<?php
}
if($pageno<>$pagecount){
?>
<a href="<?php echo $href;?>&pageno=<?php echo $pageno+1;?>">下一页</a>
<a href="<?php echo $href;?>&pageno=<?php echo $pagecount;?>">尾页</a>
<?php
}
?>
[共找到<?php echo $rows;?>个记录]
<div align="center">
<a href="student_select2_form.php">[返回]</a></div>
</div>
</body>
</html>
```

学生详情显示页面,student_detail2.php代码:

```php
<html>
<head>
<meta http-equiv="Content-Type" content="text/html; charset=gb2312">
<title>学生详细信息</title>
</head>
<body>
<?php
    $xh=$_GET["xh"];
    $link=new mysqli("127.0.0.1","root","123456");
    if(mysqli_connect_errno()){
        echo "数据库服务器连接失败!<BR>";
```

```php
        die();
    }
    $link->select_db("test") or die("数据库选择失败！<BR>");
    $link->query("set names 'gbk'");
    //执行插入操作并将结果保存在一个变量中
    $sql="select 学号,姓名,性别,班级名称,出生日期,籍贯,电话,QQ邮箱,备注,照片 from 学生,班级";
    $sql=$sql." where 学生.班级代码=班级.班级代码 and 学生.学号='$xh'";
    $result=mysqli_query($link,$sql);
//  $result = $link->query("INSERT INTO employees (picname,picture) VALUES ('myFirst','$buf')");
    //获取影响的行数
    $row=$result->fetch_object()
?>
<div align="center"><strong>学生详细信息</strong></div>
<br>
<table width="350" border="1" align="center">
  <tr>
    <td width="100"><div align="right">学号:</div></td>
    <td><? php echo $row->学号;?><div align="left"></div></td>
  </tr>
  <tr>
    <td><div align="right">姓名:</div></td>
    <td><? php echo $row->姓名;?><div align="left"></div></td>
  </tr>
  <tr>
    <td><div align="right">性别:</div></td>
    <td><? php echo $row->性别;?><div align="left"></div></td>
  </tr>
  <tr>
    <td><div align="right">班级名称:</div></td>
    <td><? php echo $row->班级名称;?><div align="left"></div></td>
  </tr>
  <tr>
    <td><div align="right">出生日期:</div></td>
    <td><? php echo $row->出生日期;?><div align="left"></div></td>
  </tr>
  <tr>
    <td><div align="right">籍贯:</div></td>
    <td><? php echo $row->籍贯;?><div align="left"></div></td>
  </tr>
  <tr>
    <td><div align="right">电话:</div></td>
    <td><? php echo $row->电话;?><div align="left"></div></td>
  </tr>
  <tr>
    <td><div align="right">QQ邮箱:</div></td>
    <td><? php echo $row->QQ邮箱;?><div align="left"></div></td>
  </tr>
  <tr>
    <td><div align="right">备注:</div></td>
    <td><? php echo $row->备注;?><div align="left"></div></td>
```

```
    </tr>
     <tr>
       <td><div align="right">照片:</div></td>
       <td><div style="width:200px; height:200px; ">
<img src="chakanzp.php? xh=<? php echo $row->学号; ? >" alt="" name="myphoto" id="myphoto" style="width:200px; height:150px">
</div>
<div align="left"></div></td>
    </tr>
</table>
<? php
  $result->free();
  $link->close();
? >
<br>
<div align="center">
<a href="student_select2_form.php">[返回]</a></div>
</body>
</html>
```

代码运行效果见图 12.19。

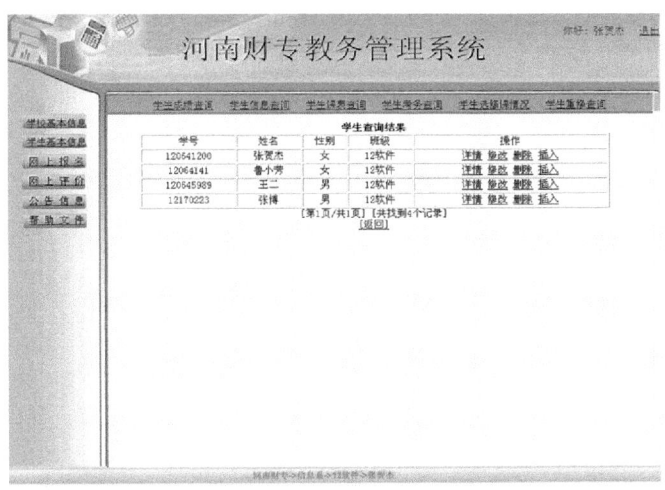

图 12.19 学生基本信息查询

➢ 学生成绩查询

查询页面,xscjcx.php 页面代码:

```
<html>
<head>
<meta http-equiv="Content-Type" content="text/html; charset=gb2312">
<title>学生查询</title>
</head>
<body><center>
<form action="xscjcx1.php" method="get">
  请选择欲查询学生所在的班级的成绩:
  <select name="bjdm" size="1">
```

```php
<?php
require("ex10_301.php");
  $link=new mysqli("127.0.0.1","root","123456");
  if (mysqli_connect_errno())  {
    echo "数据库服务器连接失败！<BR>";
    die();
  }
  $link->select_db("test") or die("数据库选择失败！<BR>");
  $link->query("set names 'gbk'");
  $sql="select 班级代码,班级名称 from 班级 order by 班级名称";
  $result=$link->query($sql);
  while( $row=$result->fetch_object())  {
?>
    <option value="<?php echo $row->班级代码;?>"><?php echo $row->班级名称;?></option>
<?php
  }
  $result->free();
  $link->close();
?>
  </select>
  <input name="submit" type="submit" value="确定">
  <input name="reset" type="reset" value="取消">
</form>
</center>
</body>
</html>
```

成绩显示页面，xscjcx1.php 代码：

```php
<html>
<head>
<meta http-equiv="Content-Type" content="text/html; charset=gb2312">
<title>学生查询结果</title>
</head>
<body>
<?php
  $bjdm=$_GET["bjdm"];
  $link=new mysqli("127.0.0.1","root","123456");
  if (mysqli_connect_errno())  {
    echo "数据库服务器连接失败！<BR>";
    die();
  }
  $link->select_db("test") or die("数据库选择失败！<BR>");
  $link->query("set names 'gbk'");
  $sql="select 学号,姓名,班级名称,课程名,成绩 from 班级,学生成绩";
  $sql.=" where 班级.班级代码=学生成绩.班级代码 ";
  $sql.=" and 班级.班级代码='$bjdm' order by 学号";
  $result=$link->query($sql);
  $rows=$result->num_rows;   //总记录数
  if ($rows==0)  {
```

```php
        echo "没有满足条件的记录!";
        die();
    }
    $pagesize=5;    //每页的记录数(在此暂设为5,通常应设为10)
    $pagecount=ceil($rows/$pagesize);    //总页数
    //$pageno的值为当前页的页号
    if(!isset($pageno)||$pageno<1)
        $pageno=1;
    if($pageno>$pagecount)
        $pageno=$pagecount;
    $offset=($pageno-1)*$pagesize;
    $result->data_seek($offset);
?>
<div align="center"><strong>学生成绩查询结果</strong></div>
<table width="90%" border="1" align="center">
    <tr>
        <td><div align="center">学号</div></td>
        <td><div align="center">姓名</div></td>
        <td><div align="center">班级名称</div></td>
        <td><div align="center">课程名</div></td>
        <td><div align="center">成绩</div></td>
        <td><div align="center">操作</div></td>
    </tr>
<?php
    $i=0;
    while($row=$result->fetch_object()) {
?>
    <tr>
        <td><div align="center"><?php echo $row->学号;?></div></td>
        <td><div align="center"><?php echo $row->姓名;?></div></td>
        <td><div align="center"><?php echo $row->班级名称;?></div></td>
        <td><div align="center"><a href="xscjcx3.php?kcm=<?php echo $row->课程名;?>" target="L2r2c1r1c22r2c1"><?php echo $row->课程名;?></a></div></td>
        <td><div align="center"><?php echo $row->成绩;?></div></td>
        <td><div align="center">
            <a href="xscjcx2.php?xh=<?php echo $row->学号;?>" target="L2r2c1r1c22r2c1">历年成绩</a>
        </div></td>
    </tr>
<?php
    $i=$i+1;
    if($i==$pagesize)
        break;
    }
    $result->free();
    $link->close();
?>
</table>
<div align="center">
[第<?php echo $pageno;?>页/共<?php echo $pagecount;?>页]
<?php
```

```php
$href=$PHP_SELF."?bjdm=".urlencode($bjdm);
if($pageno<>1){
?>
  <a href="<?php echo $href;?>&pageno=1">首页</a>
  <a href="<?php echo $href;?>&pageno=<?php echo $pageno-1;?>">上一页</a>
  <a href="xscjcx.php">[返回]</a>
<?php
}
if($pageno<>$pagecount){
?>
<a href="<?php echo $href;?>&pageno=<?php echo $pageno+1;?>">下一页</a>
<a href="<?php echo $href;?>&pageno=<?php echo $pagecount;?>">尾页</a>
<a href="xscjcx.php">[返回]</a>
<?php
}
?>
[共找到<?php echo $rows;?>个记录]
</div>
</body>
</html>
```

> 历年成绩查询

查询页面，xscjcx2.php 代码：

```php
<html>
<head>
<meta http-equiv="Content-Type" content="text/html; charset=gb2312">
<title>学生详细信息</title>
</head>
<body>
<?php
$xh=trim($xh);
$link=new mysqli("127.0.0.1","root","123456");
if(mysqli_connect_errno()){
    echo "数据库服务器连接失败！<BR>";
    die();
}
$link->select_db("test") or die("数据库选择失败！<BR>");
$link->query("set names 'gbk'");
$sql="select 学号,姓名,班级名称,课程名,成绩,学年,学期 from 班级,学生成绩";
$sql=$sql." where 班级.班级代码=学生成绩.班级代码 and 学生成绩.学号='$xh'";
$result=$link->query($sql);
    $rows=$result->num_rows;  //总记录数
if($rows==0) {
  echo "没有满足条件的记录！";
  die();
}
$pagesize=5;  //每页的记录数(在此暂设为5,通常应设为10)
$pagecount=ceil($rows/$pagesize);  //总页数
//$pageno的值为当前页的页号
if(!isset($pageno)||$pageno<1)
```

```php
    $pageno=1;
  if($pageno>$pagecount)
    $pageno=$pagecount;
  $offset=($pageno-1)*$pagesize;
  $result->data_seek($offset);
?>
<div align="center"><strong>学生成绩查询结果</strong></div>
<table width="90%" border="1" align="center">
  <tr>
    <td><div align="center">学号</div></td>
    <td><div align="center">姓名</div></td>
    <td><div align="center">班级名称</div></td>
    <td><div align="center">课程名</div></td>
    <td><div align="center">成绩</div></td>
    <td><div align="center">学年</div></td>
    <td><div align="center">学期</div></td>
  </tr>
<?php
  $i=0;
  while($row=$result->fetch_object()) {
?>
  <tr>
    <td><div align="center"><?php echo $row->学号;?></div></td>
    <td><div align="center"><?php echo $row->姓名;?></div></td>
    <td><div align="center"><?php echo $row->班级名称;?></div></td>
    <td><div align="center"><?php echo $row->课程名;?></div></td>
    <td><div align="center"><?php echo $row->成绩;?></div></td>
    <td><div align="center"><?php echo $row->学年;?></div></td>
    <td><div align="center"><?php echo $row->学期;?></div></td>
  </tr>
<?php
  $i=$i+1;
  if($i==$pagesize)
    break;
  }
  $result->free();
  $link->close();
?>
</table>
<div align="center">
[第<?php echo $pageno;?>页/共<?php echo $pagecount;?>页]
<?php
$href=$PHP_SELF."?xh=".urlencode($xh);
if($pageno<>1) {
?>
  <a href="<?php echo $href;?>&pageno=1">首页</a>
  <a href="<?php echo $href;?>&pageno=<?php echo $pageno-1;?>">上一页</a>
<?php
}
if($pageno<>$pagecount) {
?>
```

```
<a href="<?php echo $href;?>&pageno=<?php echo $pageno+1;?>">下一页</a>
<a href="<?php echo $href;?>&pageno=<?php echo $pagecount;?>">尾页</a>
<?php
}
?>
[共找到<?php echo $rows;?>个记录]
<div align="center">
<a href="xscjcx.php">[返回]</a></div>
</body>
</html>
```

代码运行效果见图12.20。

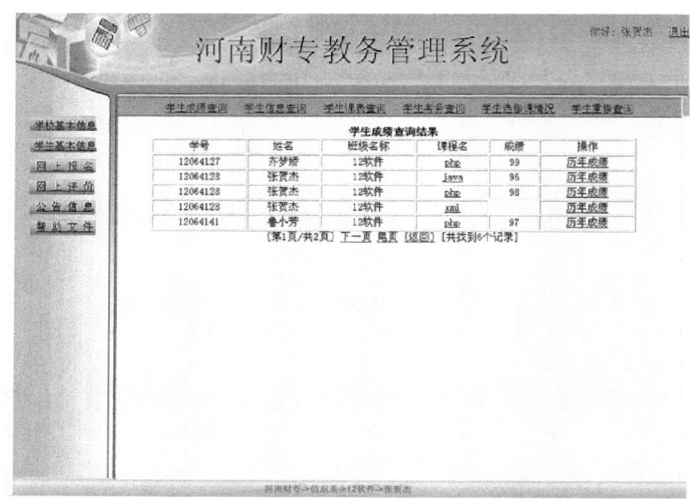

图12.20 学生成绩查询

➢ 学生课表查询

查询页面,xskb.php 页面代码:

```
<html>
<head>
<meta http-equiv="Content-Type" content="text/html; charset=gb2312">
<title>学生查询</title>
</head>
<body><center>
<form action="xskb2.php" method="get">
请选择欲查询学生所在的班级:
<select name="bjdm" size="1">
<?php
  $link=new mysqli("127.0.0.1","root","123456");
  if(mysqli_connect_errno()){
    echo "数据库服务器连接失败!<BR>";
    die();
  }
  $link->select_db("test") or die("数据库选择失败!<BR>");
  $link->query("set names 'gbk'");
  $sql="select 班级代码,班级名称 from 班级 order by 班级名称";
```

```php
    $result=$link->query($sql);
    while($row=$result->fetch_object())  {
?>
    <option value="<?php echo $row->班级代码;?>"><?php echo $row->班级名称;?></option>
<?php
    }
    $result->free();
    $link->close();
?>
    </select>
    <input name="submit" type="submit" value="确定">
    <input name="reset" type="reset" value="取消">
</form>
</center>
</body>
</html>
```

课表显示页面，xskb2.php 代码：

```php
<html>
<head>
<meta http-equiv="Content-Type" content="text/html; charset=gb2312">
<title>学生详细信息</title>
</head>
<body>
<?php
    $bjdm=$_GET["bjdm"];
    if($bjdm==""){
        echo "学号不能为空!";
        die();
    }
    $link=new mysqli("127.0.0.1","root","123456");
    if(mysqli_connect_errno()){
        echo "数据库服务器连接失败！<BR>";
        die();
    }
    $link->select_db("test")  or die("数据库选择失败！<BR>");
    $link->query("set names 'gbk'");
?>
<div align="center"><strong>学生班级课表查询结果</strong> </div>
<table width="90%" border="1" align="center">
<?php
for($row=0;$row<5;$row++){
?>
    <?php
    if($row==0){
    ?>
        <tr>
            <td><div align="center"> </div></td>
            <td><div align="center">星期一</div></td>
```

```php
            <td><div align="center">星期二</div></td>
            <td><div align="center">星期三</div></td>
            <td><div align="center">星期四</div></td>
            <td><div align="center">星期五</div></td>
        </tr>
            <?php
}else{
?>
            <tr>
                <?php
for($line=0;$line<6;$line++){
    if($row==1){
        if($line==0){
?>
            <td rowspan="2"><div align="center">上午</div></td>
<?php
        }else{
?>
            <td><div align="center">
                <?php
            $sql="select '课程名',上课地点,'姓名' from '班级课表' where '班级代码'='$bjdm' and 星期='$line' and 节次='$row'";
            $result=$link->query($sql);
            $rows=$result->num_rows;   //总记录数
            if ($rows==0)   {
?>
            <font color="#00FF00">
            <?php echo " ";?>
            </font>
            <?php
            }
            else{
while($row1=$result->fetch_object())   {
            $kcm=$row1->课程名;
            $xm=$row1->姓名;
            $skdd=$row1->上课地点;
            echo $kcm;?>
            <br>
            <?php
            echo $xm;?>
            <br><?php
            echo $skdd;
            }
            }
?>
            </div></td>
                <?php
        }
    }else if($row<3){
        if($line>0){
```

```php
                                        ?>
                                        <td><div align="center"><?php
                                        $sql="select '课程名',上课地点,'姓名' from '班级课表' where '班级代码'='$bjdm' and 星期='$line' and 节次='$row'";
                                        $result=$link->query($sql);
                                        $rows=$result->num_rows;    //总记录数
                                        if($rows==0) {
                                        ?>
                                            <font color="#00FF00">
                                            <?php echo " ";?>
                                            </font>
                                            <?php
                                        }
                                        else{
                                            while($row1=$result->fetch_object()) {
                                                $kcm=$row1->课程名;
                                                $xm=$row1->姓名;
                                                $skdd=$row1->上课地点;
                                                echo $kcm;?>
                                                <br>
                                                <?php
                                                echo $xm;?>
                                                <br><?php
                                                echo $skdd;
                                            }
                                        }
                                        ?></div></td>
                                        <?php
                                    }
                                }else if($row==3){
                                    if($line==0){
                                    ?>
                                        <td rowspan="2"><div align="center">下午</div></td>
                                    <?php
                                    }else{
                                    ?>
                                        <td><div align="center"><?php
                                        $sql="select '课程名',上课地点,'姓名' from '班级课表' where '班级代码'='$bjdm' and 星期='$line' and 节次='$row'";
                                        $result=$link->query($sql);
                                        $rows=$result->num_rows;    //总记录数
                                        if($rows==0) {
                                        ?>
                                            <font color="#00FF00">
                                            <?php echo " ";?>
                                            </font>
                                            <?php
                                        }
                                        else{
                                            while($row1=$result->fetch_object()) {
                                                $kcm=$row1->课程名;
```

```php
                $xm=$row1->姓名;
                $skdd=$row1->上课地点;
                echo $kcm;?>
                <br>
                <?php
                echo $xm;?>
                <br><?php
                echo $skdd;
                }
                }
                ?>
        </div></td>
                <?php
                }
            }else if($row>3){
                if($line>0){
                ?>
                <td><div align="center"><?php
                $sql="select '课程名',上课地点,'姓名' from '班级课表' where '班级代码'='$bjdm' and 星期='$line' and 节次='$row'";
                $result=$link->query($sql);
                $rows=$result->num_rows;   //总记录数
                if ($rows==0)  {
                                ?>
                <font color="#00FF00">
                <?php echo " ";?>
                </font>
                <?php
                }
                else{
    while($row1=$result->fetch_object())  {
                $kcm=$row1->课程名;
                $xm=$row1->姓名;
                $skdd=$row1->上课地点;
                echo $kcm;?>
                <br>
                <?php
                echo $xm;?>
                <br><?php
                echo $skdd;
                }
                }
                ?></div></td>
                <?php
                }
            }
        }
        ?>
                </tr>
                <?php
    }
```

```
}?>
</table>
<div align="center"><input name="确定" type="button" value="确定">
<a href="xskb.php">[返回]</a></div>
</body>
</html>
```

代码运行效果见图 12.21。

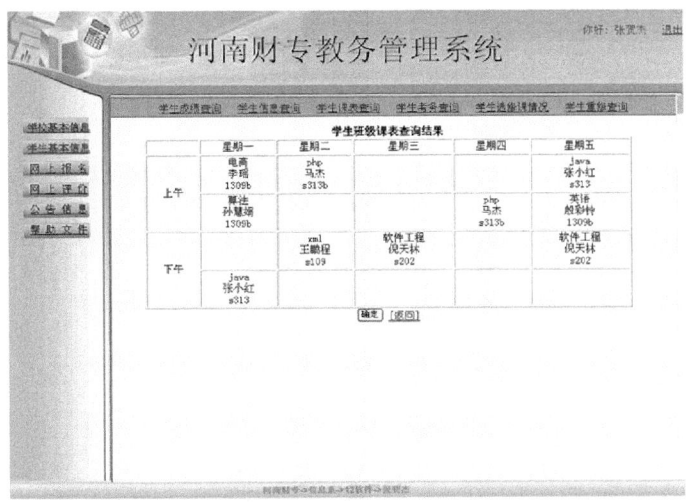

图 12.21　学生课程表

➢ 学生考务查询

查询页面,xskwcx1.php 代码:

```
<html>
<head>
<meta http-equiv="Content-Type" content="text/html; charset=gb2312">
<title>班级检索</title>
</head>
<body><center>
<form action="xskwcx.php" method="get">
请输入欲检索学生的学号:
<input name="xh" type="text" id="xh" size="9" maxlength="9">
<input name="submit" type="submit" value="确定">
<input name="reset" type="reset" value="取消">
</form></center>
</body>
</html>
```

学生考务查询结果页面,xskwcx.php 代码:

```
<html>
<head>
<meta http-equiv="Content-Type" content="text/html; charset=gb2312">
<title>学生详细信息</title>
</head>
<body>
```

```php
<?php
    $xh=$_GET["xh"];
if($xh==""){
    echo "学号不能为空!";
    die();
}
$link=new mysqli("127.0.0.1","root","123456");
if(mysqli_connect_errno()){
    echo "数据库服务器连接失败!<BR>";
    die();
}
$link->select_db("test")  or die("数据库选择失败!<BR>");
$link->query("set names 'gbk'");
$sql="select 课程名,学年,学期,考试时间,考试地点,座位号 from 学生,学生考务";
$sql=$sql." where 学生.班级代码=学生考务.班级代码 and '学生'.学号='$xh'";
$result=$link->query($sql);
 $rows=$result->num_rows;   //总记录数
if($rows==0){
    echo "没有满足条件的记录!";
    die();
}
$pagesize=5;   //每页的记录数(在此暂设为5,通常应设为10)
$pagecount=ceil($rows/$pagesize);   //总页数
//$pageno 的值为当前页的页号
if(!isset($pageno)||$pageno<1)
    $pageno=1;
if($pageno>$pagecount)
    $pageno=$pagecount;
$offset=($pageno-1)*$pagesize;
$result->data_seek($offset);
?>
<div align="center"><strong>学生考务查询结果</strong> </div>
<table width="90%" border="1" align="center">
    <tr>
        <td><div align="center">课程名</div></td>
        <td><div align="center">学年</div></td>
        <td><div align="center">学期</div></td>
        <td><div align="center">考试时间</div></td>
        <td><div align="center">考试地点</div></td>
        <td><div align="center">座位号</div></td>
    </tr>
<?php
    $i=0;
    while($row=$result->fetch_object()){
?>
    <tr>
        <td><div align="center"><?php echo $row->课程名;?></div></td>
        <td><div align="center"><?php echo $row->学年;?></div></td>
        <td><div align="center"><?php echo $row->学期;?></div></td>
        <td><div align="center"><?php echo $row->考试时间;?></div></td>
        <td><div align="center"><?php echo $row->考试地点;?></div></td>
```

```
        <td><div align="center"><?php echo $row->座位号;?></div></td>
     </tr>
<?php
   $i=$i+1;
   if($i==$pagesize)
      break;
}
$result->free();
$link->close();
?>
</table>
<div align="center">
[第<?php echo $pageno;?>页/共<?php echo $pagecount;?>页]
<?php
$href=$PHP_SELF."?xh=".urlencode($xh);
if($pageno<>1){
?>
   <a href="<?php echo $href;?>&pageno=1">首页</a>
   <a href="<?php echo $href;?>&pageno=<?php echo $pageno-1;?>">上一页</a>
<?php
}
if($pageno<>$pagecount){
?>
<a href="<?php echo $href;?>&pageno=<?php echo $pageno+1;?>">下一页</a>
<a href="<?php echo $href;?>&pageno=<?php echo $pagecount;?>">尾页</a>
<?php
}
?>
[共找到<?php echo $rows;?>个记录]
<div align="center">
<a href="xskwcx1.php">[返回]</a></div>
</body>
</html>
```

代码运行效果见图12.22。

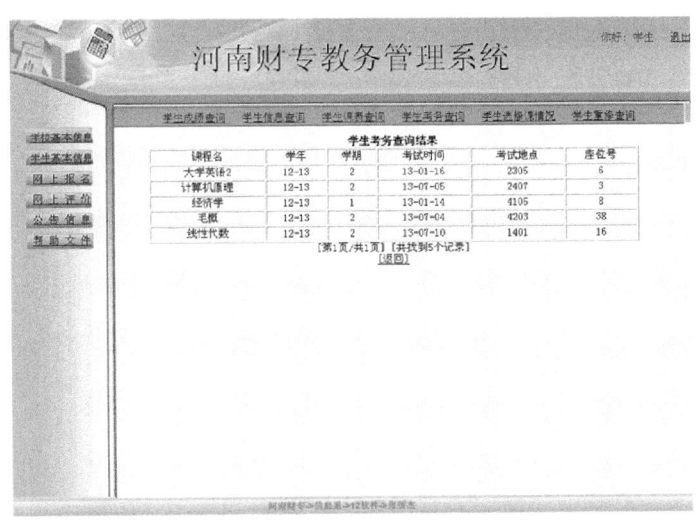

图12.22 考务查询

➤ 学生选修课查询

查询页面，xsxxkcx.php 代码：

```html
<html>
<head>
<meta http-equiv="Content-Type" content="text/html; charset=gb2312">
<title>班级检索</title>
</head>
<body><center>
<form action="xsxxkcx1.php" method="get">
    请输入欲查询学生的学号：
    <input name="xh" type="text" id="xh" size="9" maxlength="9">
    <input name="submit" type="submit" value="确定">
    <input name="reset" type="reset" value="取消">
</form></center>
</body>
</html>
```

选修课查询结果页面，xsxxkcx1.php 代码：

```php
<html>
<head>
<meta http-equiv="Content-Type" content="text/html; charset=gb2312">
<title>学生详细信息</title>
</head>
<body>
<?php
    $xh=$_GET["xh"];
if($xh=="")   {
        echo "学号不能为空!";
        die();
}
$link=new mysqli("127.0.0.1","root","123456");
if(mysqli_connect_errno()){
    echo "数据库服务器连接失败！<BR>";
    die();
}
$link->select_db("test")   or die("数据库选择失败！<BR>");
$link->query("set names 'gbk'");
$sql="select 课程名,课程号,上课地点,上课时间,任课教师,学年,学期 from 学生,学生选修课";
$sql=$sql." where 学生.姓名=学生选修课.姓名 and '学生'.学号='$xh'";
$result=$link->query($sql);
    $rows=$result->num_rows;   //总记录数
if($rows==0)   {
    echo "没有满足条件的记录!";
    die();
}
$pagesize=5;   //每页的记录数(在此暂设为5,通常应设为10)
$pagecount=ceil($rows/$pagesize);   //总页数
//$pageno 的值为当前页的页号
if(!isset($pageno)||$pageno<1)
    $pageno=1;
```

```php
    if ($pageno > $pagecount)
      $pageno = $pagecount;
    $offset = ($pageno - 1) * $pagesize;
    $result->data_seek($offset);
?>
<div align="center"><strong>学生选修课查询结果</strong></div>
<table width="90%" border="1" align="center">
  <tr>
    <td><div align="center">课程名</div></td>
    <td><div align="center">课程号</div></td>
    <td><div align="center">上课时间</div></td>
    <td><div align="center">上课地点</div></td>
    <td><div align="center">任课教师</div></td>
    <td><div align="center">学年</div></td>
    <td><div align="center">学期</div></td>
  </tr>
<?php
  $i = 0;
  while ($row = $result->fetch_object()) {
?>
  <tr>
    <td><div align="center"><?php echo $row->课程名; ?></div></td>
    <td><div align="center"><?php echo $row->课程号; ?></div></td>
    <td><div align="center"><?php echo $row->上课时间; ?></div></td>
    <td><div align="center"><?php echo $row->上课地点; ?></div></td>
    <td><div align="center"><?php echo $row->任课教师; ?></div></td>
    <td><div align="center"><?php echo $row->学年; ?></div></td>
    <td><div align="center"><?php echo $row->学期; ?></div></td>
  </tr>
<?php
    $i = $i + 1;
    if ($i == $pagesize)
      break;
  }
  $result->free();
  $link->close();
?>
</table>
<div align="center">
[第<?php echo $pageno; ?>页/共<?php echo $pagecount; ?>页]
<?php
$href = $PHP_SELF."?xh=".urlencode($xh);
if ($pageno <> 1) {
?>
  <a href="<?php echo $href; ?>&pageno=1">首页</a>
  <a href="<?php echo $href; ?>&pageno=<?php echo $pageno - 1; ?>">上一页</a>
<?php
}
if ($pageno <> $pagecount) {
?>
  <a href="<?php echo $href; ?>&pageno=<?php echo $pageno + 1; ?>">下一页</a>
```

```
<a href="<?php echo $href;?>&pageno=<?php echo $pagecount;?>">尾页</a>
<?php
}
?>
[共找到<?php echo $rows;?>个记录]
<div align="center">
<a href="xsxxkcx.php">[返回]</a></div>
</body>
</html>
```

代码运行效果见图12.23。

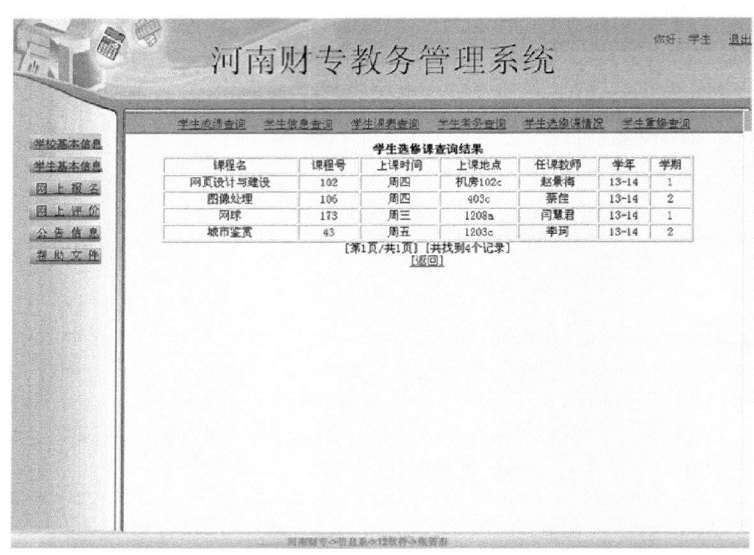

图12.23 选修课查询

➢ 学生重修查询

查询页面,xscxcx.php代码:

```
<html>
<head>
<meta http-equiv="Content-Type" content="text/html; charset=gb2312">
<title>班级检索</title>
</head>
<body><center>
<form action="xscxcx1.php" method="get">
请输入欲查询学生的学号:
<input name="xh" type="text" id="xh" size="9" maxlength="9">
<input name="submit" type="submit" value="确定">
<input name="reset" type="reset" value="取消">
</form></center>
</body>
</html>
```

查询结果页面,xscxcx1.php代码:

```
<html>
```

```php
<head>
<meta http-equiv="Content-Type" content="text/html; charset=gb2312">
<title>学生详细信息</title>
</head>
<body>
<?php
    $xh=$_GET["xh"];
  if($xh=="") {
            echo "学号不能为空!";
            die();
    }
  $link=new mysqli("127.0.0.1","root","123456");
  if(mysqli_connect_errno()){
    echo "数据库服务器连接失败!<BR>";
    die();
  }
  $link->select_db("test") or die("数据库选择失败!<BR>");
  $link->query("set names 'gbk'");
  $sql="select 姓名,性别,班级名称,课程名,学年,学期,重修成绩 from 学生,学生重修";
  $sql=$sql." where 学生.'班级代码'=学生重修.'班级代码' and '学生'.学号='$xh'";
  $result=$link->query($sql);
    $rows=$result->num_rows;   //总记录数
  if($rows==0) {
    echo "没有满足条件的记录!";
    die();
  }
  $pagesize=5;   //每页的记录数(在此暂设为5,通常应设为10)
  $pagecount=ceil($rows/$pagesize);   //总页数
  //$pageno 的值为当前页的页号
  if(!isset($pageno)||$pageno<1)
    $pageno=1;
  if($pageno>$pagecount)
    $pageno=$pagecount;
  $offset=($pageno-1)*$pagesize;
  $result->data_seek($offset);
?>
<div align="center"><strong>学生重修查询结果</strong> </div>
<table width="90%" border="1" align="center">
  <tr>
    <td><div align="center">姓名</div></td>
    <td><div align="center">性别</div></td>
    <td><div align="center">班级名称</div></td>
    <td><div align="center">课程名</div></td>
    <td><div align="center">学年</div></td>
    <td><div align="center">学期</div></td>
    <td><div align="center">重修成绩</div></td>
  </tr>
<?php
  $i=0;
  while($row=$result->fetch_object()) {
?>
```

```php
<tr>
    <td><div align="center"><?php echo $row->姓名;?></div></td>
     <td><div align="center"><?php echo $row->性别;?></div></td>
      <td><div align="center"><?php echo $row->班级名称;?></div></td>
      <td><div align="center"><?php echo $row->课程名;?></div></td>
      <td><div align="center"><?php echo $row->学年;?></div></td>
      <td><div align="center"><?php echo $row->学期;?></div></td>
      <td><div align="center"><?php echo $row->重修成绩;?></div></td>
 </tr>
<?php
  $i=$i+1;
  if($i==$pagesize)
    break;
  }
  $result->free();
  $link->close();
?>
</table>
<div align="center">
[第<?php echo $pageno;?>页/共<?php echo $pagecount;?>页]
<?php
$href=$PHP_SELF."?xh=".urlencode($xh);
if($pageno<>1){
?>
  <a href="<?php echo $href;?>&pageno=1">首页</a>
  <a href="<?php echo $href;?>&pageno=<?php echo $pageno-1;?>">上一页</a>
<?php
}
if($pageno<>$pagecount){
?>
  <a href="<?php echo $href;?>&pageno=<?php echo $pageno+1;?>">下一页</a>
  <a href="<?php echo $href;?>&pageno=<?php echo $pagecount;?>">尾页</a>
<?php
}
?>
[共找到<?php echo $rows;?>个记录]
<div align="center">
<a href="xscxcx.php">[返回]</a></div>
</body>
</html>
```

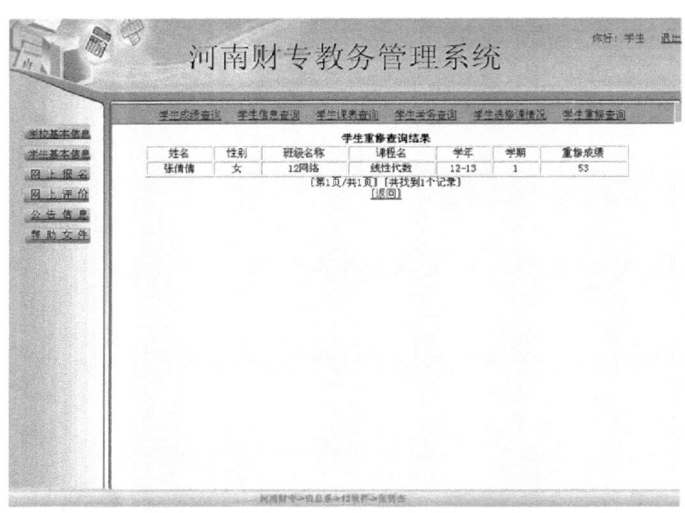

图 12.24 重修查询

代码运行效果见图 12.24。

任务六　教师功能页面

此部分包括教师信息查询和成绩录入功能，教师登录后页面和教师基本信息查询不再详述，参考学生相应页面。

➢ 成绩录入

选择录入页面，xscj_insert1.php 代码：

```php
<html>
<head>
<meta http-equiv="Content-Type" content="text/html; charset=gb2312">
<title>插入学生成绩</title>
</head>
<body><center>
<form action="xscj_insert2.php" method="get">
选择班级：
<select name="bjdm" size="1">
<?php
  $link=new mysqli("127.0.0.1","root","123456");
  if(mysqli_connect_errno()) {
     echo "数据库服务器连接失败！<BR>";
     die();
  }
  $link->select_db("test") or die("数据库选择失败！<BR>");
  $link->query("set names 'gbk'");
  $sql="select 班级代码,班级名称 from 班级 order by 班级名称";
  $result=$link->query($sql);
  while($row=$result->fetch_object()) {
?>
    <option value="<?php echo $row->班级代码;?>"><?php echo $row->班级名称;?></option>
<?php
  }
  $result->free();
  $link->close();
?>
</select>
选择课程名　<select name="kcm" size="1">
<?php
  $link=new mysqli("127.0.0.1","root","123456");
  if(mysqli_connect_errno()) {
     echo "数据库服务器连接失败！<BR>";
     die();
  }
  $link->select_db("test") or die("数据库选择失败！<BR>");
  $link->query("set names 'gbk'");
  $sql="select 课程名 from 学生课表 order by 课程名";
  $result=$link->query($sql);
  while($row=$result->fetch_object()) {
?>
```

```php
        <option value="<?php echo $row->课程名;?>"><?php echo $row->课程名;?></option>
     <?php
       }
       $result->free();
       $link->close();
      ?>
      </select>
      <label>输入学年</label><select name="xn"><option name="xn">13-14</option>
      <option name="xn">12-13</option>
      <option name="xn">11-12</option>
      <option name="xn">10-11</option>
      </select>
      <label>输入学期</label><select name="xq"><option name="xq">1</option>
      <option name="xq">2</option>
      </select>
      <input name="submit" type="submit" value="确定">
      <input name="reset" type="reset" value="取消">
    </form>
  </center>
  </body>
</html>
```

成绩录入显示页面,xscj_insert2.php 代码:

```php
<html>
<head>
<meta http-equiv="Content-Type" content="text/html; charset=gb2312">
<title>学生查询结果</title>
</head>
<body>
<form action="xscj_insert3.php" method="get">
<?php
  $bjdm=$_GET["bjdm"];
  $kcm=$_GET["kcm"];
  $link=new mysqli("127.0.0.1","root","123456");
  if (mysqli_connect_errno())  {
    echo "数据库服务器连接失败!<BR>";
    die();
  }
  $link->select_db("test") or die("数据库选择失败!<BR>");
  $link->query("set names 'gbk'");
  $sql="select 学号,姓名,课程名 from 学生成绩,班级 where 学生成绩.'班级代码'=班级.班级代码 and 学生成绩.'班级代码'=$bjdm and 学生成绩.课程名='$kcm' order by 学号";
  $result=$link->query($sql);
  $rows=$result->num_rows;   //总记录数
  if ($rows==0)  {
    echo "没有满足条件的记录!";
    die();
  }
  $pagesize=5;   //每页的记录数(在此暂设为5,通常应设为10)
```

```php
    $pagecount=ceil($rows/$pagesize);    //总页数
    //$pageno 的值为当前页的页号
    if(!isset($pageno)||$pageno<1)
       $pageno=1;
    if($pageno>$pagecount)
       $pageno=$pagecount;
    $offset=($pageno-1)*$pagesize;
    $result->data_seek($offset);
?>
<div align="center"><strong><?php echo $xn;?>学年第<?php echo $xq;?>学期插入学生成绩</strong> </div>
<table width="90%" border="1" align="center">
   <tr>
      <td><div align="center">学号</div></td>
      <td><div align="center">姓名</div></td>
       <td><div align="center">课程名 </div></td>
      <td><div align="center">成绩 </div></td>
   </tr>
<?php
   $i=0;
   while($row=$result->fetch_object()) {
?>
    <tr>
       <td><div align="center"><?php echo $row->学号;?><input type="hidden" name="xh[]" value="<?php echo $row->学号;?>"></div></td>
       <td><div align="center"><?php echo $row->姓名;?></div></td>
       <td><div align="center"><?php echo $row->课程名;?><input type="hidden" name="kcm[]" value="<?php echo $row->课程名;?>"></div></td>
       <td><div align="center"><input name="cj[]" type="text" size="10" maxlength="25"></div></td>
    </tr>
<?php
   $i=$i+1;
   if($i==$pagesize)
      break;
   }
   $result->free();
   $link->close();
?>
</table>
<div align="center">
[第<?php echo $pageno;?>页/共<?php echo $pagecount;?>页]
<?php
$href=$PHP_SELF."?bjdm=".urlencode($bjdm);
if($pageno<>1) {
?>
   <a href="<?php echo $href;?>&pageno=1">首页</a>
   <a href="<?php echo $href;?>&pageno=<?php echo $pageno-1;?>">上一页</a>
<?php
}
if($pageno<>$pagecount) {
```

```
?>
<a href="<?php echo $href;?>&pageno=<?php echo $pageno+1;?>">下一页</a>
<a href="<?php echo $href;?>&pageno=<?php echo $pagecount;?>">尾页</a>
<?php
}
?>
[共找到<?php echo $rows;?>个记录]
<div align="center">
</div>
<input name="submit" type="submit" value="确定">
</form>
</body>
</html>
```

成绩录入处理页面,xscj_insert3.php 代码:

```
<?php
for($i=0;$i<count($_GET["xh"]);$i++){
    echo $xh=$_GET["xh"][$i]." ";
    echo $kcm=$_GET["kcm"][$i]." ";
    echo $cj=$_GET["cj"][$i]." ";
    $link=mysql_connect("127.0.0.1","root","123456")
    or die("数据库服务器连接失败!<BR>");
    mysql_select_db("test",$link) or die("数据库选择失败!<BR>");
    mysql_query("set names 'gbk'");
    $sql="update 学生成绩 set 成绩='$cj' where 学号='$xh' and 课程名='$kcm'";
    if (mysql_query($sql,$link))
        echo "增加成功!";
    else
        echo '增加失败!';

}
?>
```

代码运行效果见图 12.25。

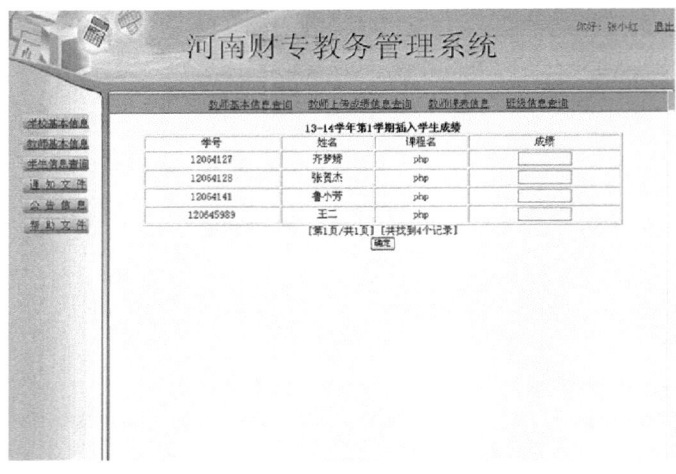

图 12.25　成绩录入

➢ 班级检索

检索页面,class_select1_form.htm 代码:

```html
<html>
<head>
<meta http-equiv="Content-Type" content="text/html; charset=gb2312">
<title>班级检索</title>
</head>
<body><center>
<form action="class_select1.php" method="get">
  请输入欲检索班级的代码:
  <input name="bjdm" type="text" id="bjdm" size="9" maxlength="9">
  <input name="submit" type="submit" value="确定">
  <input name="reset" type="reset" value="取消">
</form></center>
</body>
</html>
```

检索结果页面,class_select1.php 代码:

```php
<html>
<head>
<meta http-equiv="Content-Type" content="text/html; charset=gb2312">
<title>学生查询结果</title>
</head>
<body>
<?php
  $bjdm=$_GET["bjdm"];
  $link=new mysqli("127.0.0.1","root","123456");
  if(mysqli_connect_errno()){
    echo "数据库服务器连接失败!<BR>";
    die();
  }
  $link->select_db("test") or die("数据库选择失败!<BR>");
  $link->query("set names 'gbk'");
  $sql="select 班级代码,班级名称,班级人数,所属院,所属系 from 班级 where 班级代码='$bjdm'";
  $result=$link->query($sql);
  $rows=$result->num_rows;   //总记录数
  if($rows==0){
    echo "没有满足条件的记录!";
    die();
  }
  $pagesize=5;   //每页的记录数(在此暂设为5,通常应设为10)
  $pagecount=ceil($rows/$pagesize);   //总页数
  //$pageno 的值为当前页的页号
  if(!isset($pageno)||$pageno<1)
    $pageno=1;
  if($pageno>$pagecount)
    $pageno=$pagecount;
  $offset=($pageno-1)*$pagesize;
  $result->data_seek($offset);
?>
```

```php
<div align="center"><strong>班级查询结果</strong></div>
<table width="90%" border="1" align="center">
  <tr>
    <td><div align="center">班级代码</div></td>
    <td><div align="center">班级名称</div></td>
    <td><div align="center">班级人数</div></td>
        <td><div align="center">所属院</div></td>
        <td><div align="center">所属系</div></td>
    <td><div align="center">操作</div></td>
  </tr>
<?php
  $i=0;
  while($row=$result->fetch_object()) {
?>
  <tr>
    <td><div align="center"><?php echo $row->班级代码;?></div></td>
    <td><div align="center"><?php echo $row->班级名称;?></div></td>
    <td><div align="center"><?php echo $row->班级人数;?></div></td>
    <td><div align="center"><?php echo $row->所属院;?></div></td>
        <td><div align="center"><?php echo $row->所属系;?></div></td>
    <td><div align="center">
      <a href="class_edit.php?bjdm=<?php echo $row->班级代码;?>" target="L2r2c1r1c22r2c1">详情</a>
      <a href="class_update_edit.php?bjdm=<?php echo $row->班级代码;?>" target="L2r2c1r1c22r2c1">修改</a>
      <a href="class_delete.php?bjdm=<?php echo $row->班级代码;?>" target="L2r2c1r1c22r2c1">删除</a>
    </div></td>
  </tr>
<?php
  $i=$i+1;
  if($i==$pagesize)
    break;
  }
  $result->free();
  $link->close();
?>
</table>
<div align="center">
[第<?php echo $pageno;?>页/共<?php echo $pagecount;?>页]
<?php
$href=$PHP_SELF."?bjdm=".urlencode($bjdm);
if($pageno<>1) {
?>
  <a href="<?php echo $href;?>&pageno=1">首页</a>
  <a href="<?php echo $href;?>&pageno=<?php echo $pageno-1;?>">上一页</a>
<?php
}
if($pageno<>$pagecount) {
?>
<a href="<?php echo $href;?>&pageno=<?php echo $pageno+1;?>">下一页</a>
```

```
<a href="<?php echo $href;?>&pageno=<?php echo $pagecount;?>">尾页</a>
<?php
}
?>
[共找到<?php echo $rows;?>个记录]
<div align="center">
<a href="class_select1_form.htm">[返回]</a></div>
</div>
</body>
</html>
```

任务七 管理员功能页面

管理员主要完成对信息的录入,修改,删除。

➢ 学生信息录入

录入页面,xszp_insert.php 代码:

```
<html>
<head>
<meta http-equiv="Content-Type" content="text/html; charset=gb2312">
<title>班级增加</title>
</head>
<body>
<form action="xszp1_insert.php" method="POST" enctype="multipart/form-data" name="mainForm" id="mainForm">
<div align="center">学生查询结果</div>
<table width="350" border="1" align="center">
<tr><td width="85" class="STYLE11">学号:</td>
<td width="199" class="STYLE11">
  <div align="center">
    <input name="xh" type="text" size="20" maxlength="30" />
  </div></td>
</tr>
<tr><td class="STYLE11"><div align="center">密码:</div></td>
    <td class="STYLE11"><div align="center">
       <input name="mm" type="text" size="20" maxlength="30" />
    </div></td>
   </tr>
<tr><td class="STYLE11"><div align="center">姓名:</div></td>
    <td class="STYLE11"><div align="center">
       <input name="xm" type="text" size="20" maxlength="30" />
    </div></td>
   </tr>
      <tr><td class="STYLE11"><div align="center">班级代码:</div></td>
    <td class="STYLE11"><div align="center">
       <input name="bjdm" type="text" size="20" maxlength="30" />
    </div></td>
       </tr>
<tr><td class="STYLE11"><div align="center">性别:</div></td>
```

```html
        <td class="STYLE11"><div align="center">
            <input name="xb" type="text" size="20" maxlength="30" />
        </div></td>
      </tr>
        <tr><td class="STYLE11"><div align="center">籍贯：</div></td>
        <td class="STYLE11"><div align="center">
            <input name="jg" type="text" size="20" maxlength="30" />
        </div></td>
       </tr>
        <tr><td class="STYLE11"><div align="center">电话：</div></td>
        <td class="STYLE11"><div align="center">
            <input name="dh" type="text" size="20" maxlength="30" />
        </div></td>
       </tr>
        <tr><td class="STYLE11"><div align="center">QQ邮箱：</div></td>
        <td class="STYLE11"><div align="center">
            <input name="qqyx" type="text" size="20" maxlength="30" />
        </div></td>
       </tr>
        <tr><td class="STYLE11"><div align="center">备注：</div></td>
        <td class="STYLE11"><div align="center">
            <input name="bz" type="text" size="20" maxlength="30" />
        </div></td>
       </tr>
        <tr><td class="STYLE11"><div align="center">照片：</div></td>
        <td class="STYLE11"><div align="center">
            <img src="" name="myphoto" /><br>
  <input type="file" name="myFile"  onchange="mainForm.myphoto.src=this.value;" />
<br />
        </div></td>
      </tr>
    </table>
</form>
<div align="center">
    <input name="submit" type="submit" value="确定">
    <input name="reset" type="reset" value="取消">
  </div>
</body>
</html>
```

录入结果处理页面,xszp1_insert.php代码：

```php
<?php
$xh=$_POST["xh"];
$mm=$_POST["mm"];
$xm=$_POST["xm"];
$bjdm=$_POST["bjdm"];
$xb=$_POST["xb"];
$jg=$_POST["jg"];
$dh=$_POST["dh"];
$qqyx=$_POST["qqyx"];
```

```
$bz=$_POST["bz"];
$photoname=$_FILES['myFile']['tmp_name'];
if(!empty($photoname)){
      $photo=fread(fopen($photoname,"r"),filesize($photoname));
      $photo = '0x'. bin2hex($photo);}
  $link=mysql_connect("127.0.0.1","root","123456")
or die("数据库服务器连接失败!");
mysql_select_db("test",$link) or die("数据库选择失败!");
mysql_query("set names 'gbk'");
  //执行插入操作并将结果保存在一个变量中
  $sql="select 学号 from 学生 where 学号='$xh'";
  $result=mysql_query($sql,$link);
  $row=mysql_fetch_array($result);
if($row){
echo "此学号已经存在!";
die();
}
$sql="insert into 学生(学号,密码,姓名,班级代码,性别,籍贯,电话,QQ邮箱,备注,照片)";
$sql=$sql."
values('$xh','$mm','$xm','$bjdm','$xb','$jg','$dh','$qqyx','$bz',$photo)";
if(mysql_query($sql,$link))
echo "学生增加成功!";
else
  echo '学生增加失败!';
?>
```

代码运行效果见图12.26。

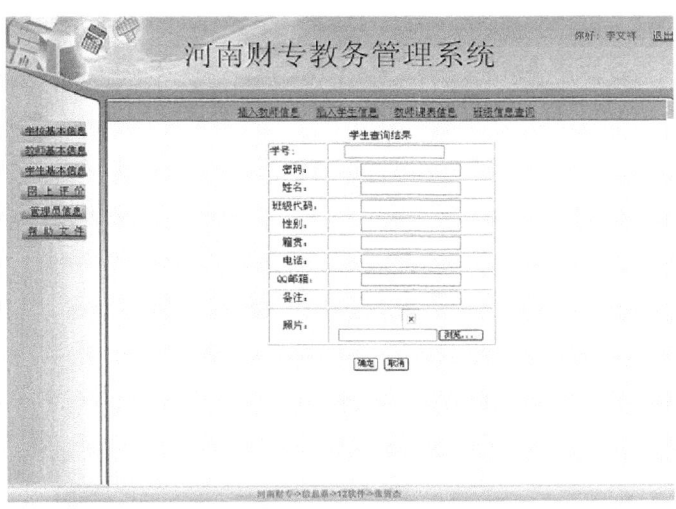

图 12.26　增加信息

> 班级查询页

查询页面,class_select1_form.htm 代码：

```
<html>
<head>
<meta http-equiv="Content-Type" content="text/html; charset=gb2312">
```

```html
<title>班级检索</title>
</head>
<body><center>
<form action="class_select1.php" method="get">
   请输入欲检索班级的代码:
   <input name="bjdm" type="text" id="bjdm" size="9" maxlength="9">
   <input name="submit" type="submit" value="确定">
   <input name="reset" type="reset" value="取消">
</form></center>
</body>
</html>
```

查询结果页面,class_select1.php 代码:

```html
<html>
<head>
<meta http-equiv="Content-Type" content="text/html; charset=gb2312">
<title>学生查询结果</title>
</head>
<body>
<?php
  $bjdm=$_GET["bjdm"];
  $link=new mysqli("127.0.0.1","root","123456");
  if(mysqli_connect_errno()) {
     echo "数据库服务器连接失败! <BR>";
     die();
  }
  $link->select_db("test") or die("数据库选择失败! <BR>");
  $link->query("set names 'gbk'");
  $sql="select 班级代码,班级名称,班级人数,所属院,所属系 from 班级 where 班级代码='$bjdm'";
  $result=$link->query($sql);
  $rows=$result->num_rows;   //总记录数
  if($rows==0) {
     echo "没有满足条件的记录!";
     die();
  }
  $pagesize=5;   //每页的记录数(在此暂设为5,通常应设为10)
  $pagecount=ceil($rows/$pagesize);   //总页数
  //$pageno 的值为当前页的页号
  if(!isset($pageno)||$pageno<1)
     $pageno=1;
  if($pageno>$pagecount)
     $pageno=$pagecount;
  $offset=($pageno-1)*$pagesize;
  $result->data_seek($offset);
?>
<div align="center"><strong>班级查询结果</strong> </div>
<table width="90%" border="1" align="center">
  <tr>
    <td><div align="center">班级代码</div></td>
    <td><div align="center">班级名称</div></td>
```

```php
        <td><div align="center">班级人数</div></td>
        <td><div align="center">所属院</div></td>
        <td><div align="center">所属系</div></td>
        <td><div align="center">操作</div></td>
    </tr>
<?php
    $i=0;
    while($row=$result->fetch_object()) {
?>
    <tr>
        <td><div align="center"><?php echo $row->班级代码;?></div></td>
        <td><div align="center"><?php echo $row->班级名称;?></div></td>
        <td><div align="center"><?php echo $row->班级人数;?></div></td>
        <td><div align="center"><?php echo $row->所属院;?></div></td>
        <td><div align="center"><?php echo $row->所属系;?></div></td>
        <td><div align="center">
            <a href="class_edit.php?bjdm=<?php echo $row->班级代码;?>" target="L2r2c1r1c22r2c1">详情</a>
            <a href="class_update_edit.php?bjdm=<?php echo $row->班级代码;?>" target="L2r2c1r1c22r2c1">修改</a>
            <a href="class_delete.php?bjdm=<?php echo $row->班级代码;?>" target="L2r2c1r1c22r2c1">删除</a>
        </div></td>
    </tr>
<?php
    $i=$i+1;
    if ($i==$pagesize)
        break;
    }
    $result->free();
    $link->close();
?>
</table>
<div align="center">
[第<?php echo $pageno;?>页/共<?php echo $pagecount;?>页]
<?php
$href=$PHP_SELF."?bjdm=".urlencode($bjdm);
if ($pageno<>1) {
?>
    <a href="<?php echo $href;?>&pageno=1">首页</a>
    <a href="<?php echo $href;?>&pageno=<?php echo $pageno-1;?>">上一页</a>
<?php
}
if ($pageno<>$pagecount) {
?>
<a href="<?php echo $href;?>&pageno=<?php echo $pageno+1;?>">下一页</a>
<a href="<?php echo $href;?>&pageno=<?php echo $pagecount;?>">尾页</a>
<?php
}
?>
[共找到<?php echo $rows;?>个记录]
```

```
<div align="center">
<a href="class_select1_form.htm">[返回]</a></div>
</div>
</body>
</html>
```

代码运行效果见图 12.27。

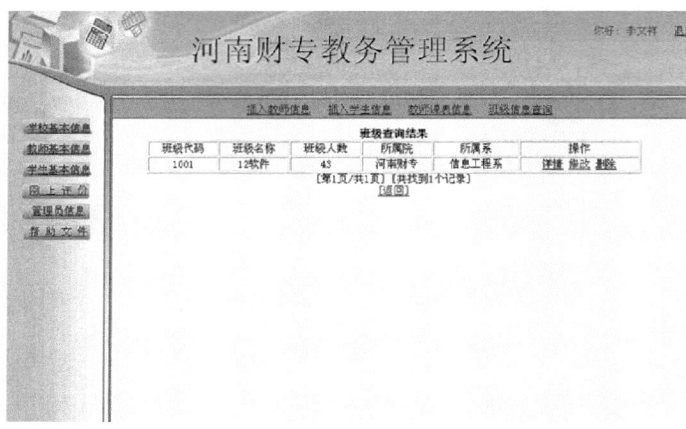

图 12.27　班级查询

➢ 信息修改

修改选择页面,class_update_edit.php 代码:

```
<html>
<head>
<meta http-equiv="Content-Type" content="text/html; charset=gb2312">
<title>班级编辑</title>
</head>
<body>
<? php
  $bjdm=$_GET["bjdm"];
  if($bjdm=="")  {
       echo "班级代码不能为空!";
       die();
  }
  $link=mysql_connect("127.0.0.1","root","123456")
      or die("数据库服务器连接失败!<BR>");
  mysql_select_db("test",$link)
or die("数据库选择失败!<BR>");
  mysql_query("set names 'gbk'");
  $sql="select 班级代码,班级名称,班级人数,所属院,所属系 from 班级 where 班级代码='$bjdm'";
  $result=mysql_query($sql,$link);
  $row = mysql_fetch_array($result);
  if(! $row)  {
     echo "无此班级代码!";
     die();
  }
  $bjdm=$row['班级代码'];
```

```php
    $bjmc = $row['班级名称'];
    $bjrs = $row['班级人数'];
    $ssy = $row['所属院'];
    $ssx = $row['所属系'];
?>
<form action="class_update.php" method="get">
  <div align="center">班级编辑</div>  <br>
  <table width="300" border="1" align="center">
  <tr><td width="85">班级代码：</td>
    <td width="199"><input name="bjdm" type="text" value="<?php echo $bjdm;?>" size="9" maxlength="9"></td>
  </tr>
  <tr>  <td>班级简称：</td>
    <td><input name="bjmc" type="text" value="<?php echo $bjmc;?>" size="15" maxlength="15"></td>
  </tr>
  <tr>  <td>班级全称：</td>
    <td><input name="bjrs" type="text" value="<?php echo $bjrs;?>" size="20" maxlength="30"></td>
  </tr>
  <tr>  <td>所属院：</td>
    <td><input name="ssy" type="text" value="<?php echo $ssy;?>" size="20" maxlength="30"></td>
  </tr>
  <tr>  <td>所属系：</td>
    <td><input name="ssx" type="text" value="<?php echo $ssx;?>" size="20" maxlength="30"></td>
  </tr>
  </table>
  <input name="bjdm0" type="hidden" value="<?php echo $bjdm;?>">
  <br>
  <div align="center">
    <input name="submit" type="submit" value="确定">
    <input name="reset" type="reset" value="取消">
  </div>
</form>
</body>
</html>
```

修改处理页面，class_update.php 代码：

```php
<?php
    $bjdm = $_GET["bjdm"];
    $bjmc = $_GET["bjmc"];
    $bjrs = $_GET["bjrs"];
    $ssy = $_GET["ssy"];
    $ssx = $_GET["ssx"];
    $bjdm0 = $_GET["bjdm0"];
    if ($bjdm=="" || $bjmc=="" || $bjrs=="") {
        echo "班级代码及其简称与全称均不能为空！";
        die();
```

```php
    }
    $link=mysql_connect("127.0.0.1","root","123456")
        or die("数据库服务器连接失败！<BR>");
mysql_select_db("test",$link) or die("数据库选择失败！<BR>");
mysql_query("set names 'gbk'");
if($bjdm!=$bjdm0){
    $sql="select 班级代码 from 班级 where 班级代码='$bjdm'";
    $result=mysql_query($sql,$link);
    $row = mysql_fetch_array($result);
    if($row){
        echo "此班级代码已经存在！";
        die();
    }
}
$sql="update 班级 set 班级代码='$bjdm',班级名称='$bjmc' ";
$sql=$sql.",班级人数='$bjrs',所属院='xxy',所属系='$ssx' where bjdm='$bjdm0'";
if(mysql_query($sql,$link))
    echo "班级修改成功！";
else
    echo '班级修改失败！';
?>
```

代码运行效果见图 12.28。

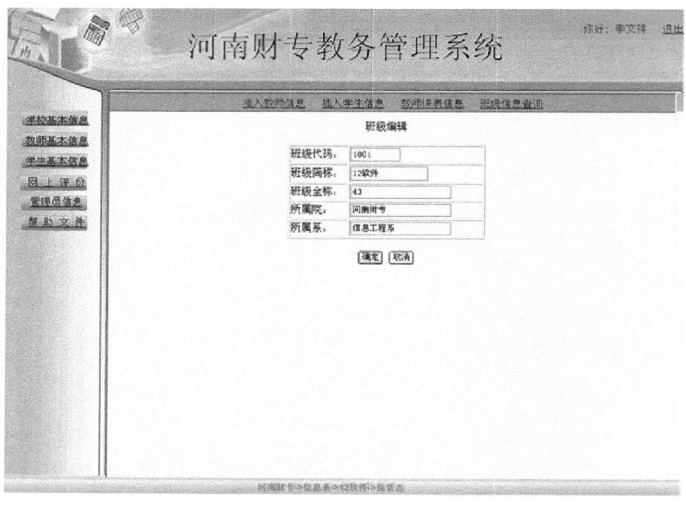

图 12.28　班级编辑

> 信息删除

删除页面，class_delete.php 代码：

```php
<?php
$bjdm=trim($bjdm);
if($bjdm==""){
    echo "班级代码不能为空！";
    die();
}
$link=mysql_connect("127.0.0.1","root","123456")
```

```php
    or die("数据库服务器连接失败！<BR>");
mysql_select_db("test", $link) or die("数据库选择失败！<BR>");
mysql_query("set names 'gbk'");
    $sql = "select bjdm from bj where bjdm='$bjdm'";
    $result = mysql_query($sql, $link);
    $row = mysql_fetch_array($result);
    if (! $row) {
        echo "无此班级代码!";
        die();
    }
    $sql = "delete from bj where bjdm='$bjdm'";
    if (mysql_query($sql, $link))
        echo "班级删除成功!";
    else
        echo '班级删除失败！';
?>
```

五、考核标准

（1）版面布局合理清晰，整体效果美观，观赏性强。（10分）
（2）网页中没有明显的错误（如超链接、图片无法显示、错别字等）。（10分）
（3）学生功能模块。（40分）
（4）教师功能模块。（20分）
（5）管理员功能模块。（10分）
（6）创新性、其他功能。（10分）

项目十三　库存管理系统

一、实训目的
- 掌握库存资产类管理系统的开发方法；
- 掌握新闻发布在网站中的应用。

二、实训要求
- 使用 PHP 开发库存管理系统；
- 将新闻公告功能加入库存管理系统。

三、实训设计

库存管理系统，它是一个典型的数据库应用程序，集进货，销售，存储多个环节于一体的信息系统。根据企业的需求，解决了企业库存不准，信息反馈不及时等一系列问题，为企业创造了良好的经济效益。

➤ 模块设计
- 登录模块：填写已分配的用户编号，正确的密码，进行用户切换，进入主控制页面。
- 销售员模块：销售员信息的查询，店铺信息的查询，商品信息的查询，商品的销售，新闻动态的查询，库存警报的查询。
- 经理模块：销售员信息的查询，增加，经理信息的查询，增加，仓库管理员信息的查询，增加，店铺信息的查询，增加，商品价格的修改，新闻动态的增加，修改，删除。
- 仓库管理员模块：商品的入库，出库，入库单的查询，导出，出库单的查询，导出，新闻动态的查询，库存警报的查询。

四、实训内容

任务一　功能模块设计

➤ 库存管理系统功能
(1) 库存管理系统是以进货，销售，库存为主要功能的系统。
(2) 销售员信息包括编号，姓名，联系电话，所属店铺号等的信息。
(3) 经理信息包括经理编号，姓名，联系电话等的信息。
(4) 仓库管理员信息包括仓库管理员号，姓名，联系电话等的信息。
(5) 商品信息包括物品名称，物品编号，颜色，大小等的信息。
(6) 公告信息包括公告 ID，标题，正文，上传时间等的信息。
(7) 库存警报信息包括物品编号，物品名称，颜色，大小，是否存在库存警报等的信息。

➤ 功能结构示意图
(1) 登录功能示意图，见图 13.1。

（2）销售员模块功能示意图，见图13.2。

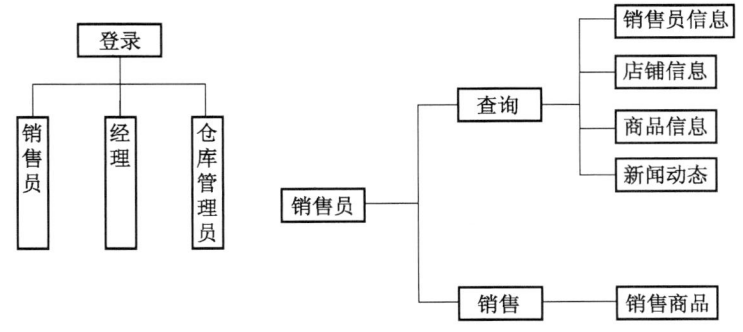

图13.1 登录功能　　　　　图13.2 销售员功能

（3）经理模块功能示意图，见图13.3。
（4）仓库管理员模块功能示意图，见图13.4。

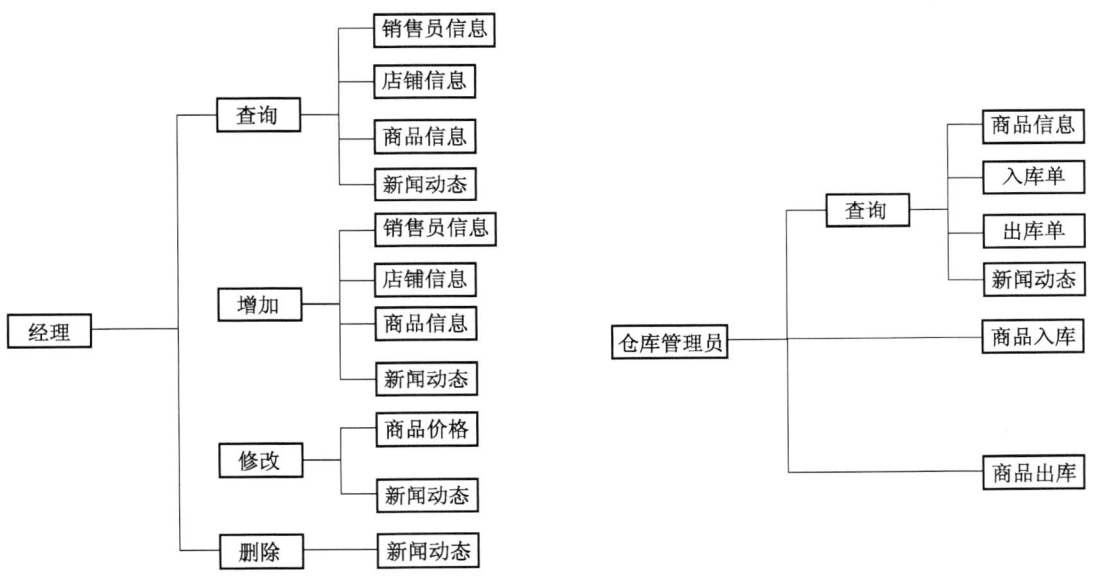

图13.3 经理功能　　　　　图13.4 仓库管理员功能

任务二　数据库设计

（1）销售员信息表，见图13.5。

名	类型	长度	小数点	允许空值(	
销售员号	varchar	12	0	☐	🔑1
销售员姓名	varchar	8	0	☐	
性别	varchar	2	0	☑	
民族	varchar	6	0	☑	
出生日期	date	0	0	☑	
联系电话	varchar	12	0	☑	
店铺号	varchar	12	0	☐	
密码	varchar	8	0	☐	

图13.5 销售员信息表

(2) 经理信息表,见图 13.6。

名	类型	长度	小数点	允许空值	
经理编号	varchar	12	0	☐	🔑1
经理姓名	varchar	8	0	☐	
性别	varchar	2	0	☑	
民族	varchar	2	0	☑	
出生日期	date	0	0	☑	
联系电话	varchar	12	0	☑	
店铺号	varchar	12	0	☐	
密码	varchar	8	0	☐	

图 13.6　经理信息表

(3) 库存管理员信息表,见图 13.7。

名	类型	长度	小数点	允许空值	
库存管理员号	varchar	12	0	☐	🔑1
库存管理员姓名	varchar	8	0	☐	
性别	varchar	2	0	☑	
民族	varchar	6	0	☑	
出生日期	date	0	0	☑	
联系电话	varchar	12	0	☑	
密码	varchar	8	0	☐	

图 13.7　库存管理员信息表

(4) 店铺信息表,见图 13.8。

名	类型	长度	小数点	允许空值	
店铺号	varchar	12	0	☐	🔑1
店名称	varchar	10	0	☐	
店铺图	blob	0	0	☑	
店铺简介	varchar	32000	0	☑	
地址	varchar	20	0	☑	
电话	varchar	11	0	☑	

图 13.8　店铺信息表

(5) 店铺物品表,见图 13.9。

名	类型	长度	小数点	允许空值	
物品编号	varchar	12	0	☐	🔑1
物品名称	varchar	12	0	☐	
颜色	varchar	8	0	☐	🔑2
大小	varchar	6	0	☐	🔑3
产地	varchar	26	0	☐	
售价	varchar	10	0	☐	
折扣	float	6	2	☑	
数量	varchar	10	0	☐	
店铺号	varchar	6	0	☐	🔑4
分类编号	varchar	8	0	☐	

图 13.9　店铺物品表

(6) 物品信息表,见图 13.10。
(7) 分类表,见图 13.11。
(8) 销售表,见图 13.12。

名	类型	长度	小数点	允许空值	
物品编号	varchar	12	0	☐	🔑1
物品名称	varchar	12	0	☑	
颜色	varchar	8	0	☐	🔑2
大小	varchar	6	0	☐	🔑3
产地	varchar	26	0	☑	
售价	varchar	10	0	☐	
折扣	float	6	2	☑	
分类编号	varchar	8	0	☑	

图 13.10　物品信息表

名	类型	长度	小数点	允许空值	
分类编号	varchar	8	0	☐	🔑1
备注	varchar	16	0	☑	

图 13.11　分类表

名	类型	长度	小数点	允许空值	
物品编号	varchar	12	0	☐	🔑1
物品名称	varchar	12	0	☐	
颜色	varchar	8	0	☐	🔑2
大小	varchar	6	0	☐	🔑3
售价	varchar	10	0	☐	
折扣	float	6	2	☑	
数量	varchar	10	0	☐	
店铺号	varchar	6	0	☐	🔑4
分类编号	varchar	8	0	☐	

图 13.12　销售表

(9) 实际库存表,见图 13.13。

名	类型	长度	小数点	允许空值	
物品编号	varchar	12	0	☐	🔑1
物品名称	varchar	20	0	☑	
颜色	varchar	8	0	☐	🔑2
大小	varchar	6	0	☐	🔑3
物品数量	varchar	6	0	☐	
进价	varchar	10	0	☐	

图 13.13　实际库存表

(10) 入库表,见图 13.14。

名	类型	长度	小数点	允许空值	
入库编号	varchar	12	0	☐	🔑1
物品编号	varchar	12	0	☐	🔑2
物品名称	varchar	20	0	☐	
颜色	varchar	8	0	☐	🔑3
大小	varchar	6	0	☐	🔑4
进价	varchar	10	0	☐	
货位编号	varchar	15	0	☐	
入库日期	date	0	0	☐	
入库数量	varchar	8	0	☐	
负责人	varchar	8	0	☑	

图 13.14　入库表

(11) 出库表,见图 13.15。

名	类型	长度	小数点	允许空值(	
出库编号	varchar	12	0	☐	🔑1
物品编号	varchar	12	0	☐	🔑2
物品名称	varchar	20	0	☐	
颜色	varchar	8	0	☐	🔑3
大小	varchar	6	0	☐	🔑4
进价	varchar	10	0	☐	
货位编号	varchar	15	0	☐	
出库日期	date	0	0	☐	
出库数量	varchar	8	0	☐	
负责人	varchar	8	0	☑	

图 13.15　出库表

(12) 公告表,见图 13.16。

名	类型	长度	小数点	允许空值(	
公告ID	varchar	20	0	☐	🔑1
标题	varchar	16	0	☐	
正文	varchar	31000	0	☐	
上传人	varchar	6	0	☐	
上传时间	date	0	0	☐	

图 13.16　公告表

任务三　登录功能

本模块主要完成对用户账号密码的校验,账号密码错误则不允许登录。
登录页面,index1.html 代码:

```html
<!DOCTYPE html>
<html lang="en" class="no-js">
    <head>
<meta http-equiv="Content-Type" content="text/html; charset=gb2312" />
        <meta charset="utf-8">
        <title>Insole Inventory</title>
        <meta name="viewport" content="width=device-width, initial-scale=1.0">
        <meta name="description" content="">
        <meta name="author" content="">
        <!-- CSS -->
        <!-- <link rel='stylesheet' href='http://fonts.googleapis.com/css?family=PT+Sans:400,700'>
        <link rel="stylesheet" href="css/reset.css">
        <link rel="stylesheet" href="css/supersized.css"> -->
        <link rel="stylesheet" href="css/style.css">
        <!-- HTML5 shim, for IE6-8 support of HTML5 elements -->
        <!--[if lt IE 9]>
            <script src="http://html5shim.googlecode.com/svn/trunk/html5.js"></script>
```

```html
        <![endif]-->
    </head>
    <body background="img/backgrounds/5.jpg">
        <div class="page-container">
            <h1>Insole Inventory</h1>
            <form action="index.php" method="get">
                <input type="text"  style="width:270px;height:40px;" name="username" placeholder="Username" ><br><br><br><br>
                <input type="password"  style="width:270px;height:40px;" name="password" placeholder="Password"><br><br>

                <input type="radio" name="sf" value="销售员"><font color=white size=4>销售员</font> 
                <input type="radio" name="sf" value="经理"><font color=white size=4>经理</font> 
                <input type="radio" name="sf" value="库存管理员"><font color=white size=4>库存管理员</font>
                <button type="submit">Sign me in</button>
                <div class="error"><span>+</span></div>
            </form>
            <!--<form action="index.php"  method="get">
<b><font color=blue size=5>登   录:</font></b>
<input type="text" name="username"><br><br><br>
<b><font color=blue size=5>密   码:</font></b>
<input type="password" name="password"><br><br><br>
<input type="radio" name="sf" value="销售员"><font color=black size=4>销售员</font>  
<input type="radio" name="sf" value="经理"><font color=black size=4>经理</font>  
<input type="radio" name="sf" value="库存管理员"><font color=black size=4>库存管理员</font><br><br><br>
<input type="submit" style=" font-size:20px;"  value="登录">    
<input type="reset" style=" font-size:20px;" value="重置">
        </form>
         -->



        </div>
        <!-- Javascript -->
        <!-- <script src="js/jquery-1.8.2.min.js"></script>
        <script src="js/supersized.3.2.7.min.js"></script>
        <script src="js/supersized-init.js"></script>
        <script src="js/scripts.js"></script>
 -->
    </body>
</html>
```

登录处理页面,index.php 代码:

```
<!DOCTYPE html PUBLIC "-//W3C//DTD XHTML 1.0 Transitional//EN" "http://www.w3.org/TR/xhtml1/DTD/xhtml1-transitional.dtd">
```

```php
<html xmlns="http://www.w3.org/1999/xhtml">
<head>
<meta http-equiv="Content-Type" content="text/html; charset=gb2312" />
<title>无标题文档</title>
</head>

<body>
<?php

  session_start();
if(!isset($_SESSION['name'])){
    if(@$_GET['sf']==""){
    echo "您未选择身份,请选择身份!";
    header("refresh:2;url=index1.html");
                exit;
    }

    if($_GET['sf']=="销售员"){
        $xh = $_GET['username'];
            $password = $_GET['password'];
            mysql_connect("127.0.0.1","root","123456");
            mysql_select_db("kcgl");
        $query ="SELECT 销售员号,销售员姓名 FROM 销售员信息表  WHERE 销售员号='$xh' AND 密码='$password'";
            mysql_query("set names 'gbk'");
        $result=mysql_query($query);
        if (mysql_num_rows($result) == 1) {
        $_SESSION['name']=mysql_result($result,0,"销售员姓名");
        $_SESSION['销售员号']=mysql_result($result,0,"销售员号");
            header("refresh:1;url=1.php");
        } else {
            echo "密码错误或者未输入,请重新输入!";
                header("refresh:2;url=index.html");
                exit;
        }
    }
    else if($_GET['sf']=="经理"){
        $jsh = $_GET['username'];
        $password = $_GET['password'];
        mysql_connect("127.0.0.1","root","123456");
        mysql_select_db("kcgl");
            $query ="SELECT 经理编号,经理姓名 FROM 经理信息表  WHERE 经理编号='$jsh' AND 密码='$password'";
            mysql_query("set names 'gbk'");
    $result=mysql_query($query);
    if (mysql_num_rows($result) == 1) {
    $_SESSION['name']=mysql_result($result,0,"经理姓名");
    $_SESSION['经理编号']=mysql_result($result,0,"经理编号");
                header("refresh:1;url=2.php");
        } else {
            echo "密码错误或者未输入,请重新输入!";
```

```
                header("refresh:2;url = index.html");
                exit;
            }
        }
    else if( $_GET['sf'] == "库存管理员"){
        $ab = $_GET['username'];
            $password = $_GET['password'];
            mysql_connect("127.0.0.1","root","123456");
            mysql_select_db("kcgl");
                $query = "SELECT 库存管理员号,库存管理员姓名 FROM 库存管理员信息表
WHERE 库存管理员号 = '$ab' AND 密码 = '$password'";
                mysql_query("set names 'gbk'");
            $result = mysql_query($query);
            if (mysql_num_rows($result) == 1) {
            $_SESSION['name'] = mysql_result($result,0,"库存管理员姓名");
            $_SESSION['库存管理员号'] = mysql_result($result,0,"库存管理员号");
                header("refresh:1;url = 3.php");
        } else {
            echo "密码错误或者未输入,请重新输入!";
                header("refresh:2;url = index.html");
                exit;
            }
        }
    }
?>
</body>
</html>
```

代码运行效果见图 13.17。

图 13.17 登录功能

任务四 店外商品查询功能

主要完成对店外商品的查询,可以按照物品编号、颜色大小进行查询。

查询页面,1_dianwai_shangpin_xx_1.php 代码:

```
<html>
<head>
```

```html
<title>销售员信息</title>
</head>
<body>
<form action="1_dianwai_shangpin_xx.php" target="right_2" method="get">
<b>物品编号:</b>
<input type="text" style="width:70px;height:20px;" name="物品编号">
<b>颜色:</b>
<input type="text" style="width:70px;height:20px;" name="颜色">
<b>大小:</b>
<input type="text" style="width:70px;height:20px;" name="大小">

<input type="submit" style="width:50px;height:30px;" value="查询">
<input type="reset" style="width:50px;height:30px;" value="取消">
</form>
</body>
</html>
```

查询结果页面，1_dianwai_shangpin_xx.php 代码：

```php
<html>
<head>
<meta http-equiv="Content-Type" content="text/html; charset=gb2312">
<title>销售员查询结果</title>
</head>
<body>
<?php
$link=mysql_connect("127.0.0.1","root","123456") or die("数据库服务器连接失败！<br>");
mysql_select_db("kcgl",$link) or die("数据库选择成功！<br>");
   mysql_query("set names 'gbk'");
if($_GET['物品编号'])
  {$物品编号=$_GET['物品编号'];
$sql="select * from 店铺物品表 where 店铺物品表.物品编号='$物品编号'";}
if($_GET['颜色'])
  {$颜色=$_GET['颜色'];
$sql="select * from 店铺物品表 where  颜色='$颜色'";}
if($_GET['大小'])
  {$大小=$_GET['大小'];
$sql="select * from 店铺物品表 where 大小='$大小'";}
if($_GET['物品编号']&&$_GET['颜色'])
  {$物品编号=$_GET['物品编号'];
   $颜色=$_GET['颜色'];
$sql="select * from 店铺物品表 where 物品编号='$物品编号' and 颜色='$颜色'";}
if($_GET['物品编号']&&$_GET['大小'])
  {$物品编号=$_GET['物品编号'];
   $大小=$_GET['大小'];
$sql="select * from 店铺物品表 where 物品编号='$物品编号' and 大小='$大小'";}
if($_GET['颜色']&&$_GET['大小'])
  {$颜色=$_GET['颜色'];
   $大小=$_GET['大小'];
$sql="select * from 店铺物品表 where 颜色='$颜色' and 大小='$大小'";}
if($_GET['物品编号']&&$_GET['颜色']&&$_GET['大小'])
```

```php
       {$物品编号=$_GET['物品编号'];
        $颜色=$_GET['颜色'];
        $大小=$_GET['大小'];
    $sql="select 物品编号,物品名称,颜色,大小,售价,折扣,数量,店铺号 from 店铺物品表 where 物品编号='$物品编号' and 颜色='$颜色' and 大小='$大小'";}
    $result=mysql_query($sql,$link) or die("有错<br>");
    $rows=mysql_num_rows($result);
      if ($rows==0)  {
        echo "没有满足条件的记录!";
        die();
      }
      $pagesize=12;  //每页的记录数(在此暂设为5,通常应设为10)
      $pagecount=ceil($rows/$pagesize);   //总页数
     //$pageno 的值为当前页的页号
      if (!isset($pageno)||$pageno<1)
         $pageno=1;
      if ($pageno>$pagecount)
         $pageno=$pagecount;
      $offset=($pageno-1)*$pagesize;
    mysql_data_seek($result,$offset);
    ?>
    <div align="center"><strong>物品查询结果</strong> </div>
    <table style="table-layout:fixed;"  width="90%" border="3" cellpadding="0" cellspacing="0"  bordercolor="#99FFFF" align="center"  bgcolor="#CDE0F1">
    <thead bgcolor="#3399FF">
      <tr>
        <td width="80"><div align="center">物品编号</div></td>
        <td  width="100"><div align="center">物品名称</div></td>
        <td width="60"><div align="center">颜色</div></td>
        <td width="60"><div align="center">大小</div></td>
        <!--<td width="90"><div align="center">产地</div></td>-->
        <td width="60"><div align="center">售价</div></td>
        <td width="60"><div align="center">折扣</div></td>
        <td width="60"><div align="center">数量</div></td>
        <td width="60"><div align="center">存放位置</div></td>
        <!--<td width="90"><div align="center">分类编号</div></td>-->
      </tr>
    </thead>
  <?php
     $i=0;
     while($row=mysql_fetch_array($result))  {
   ?>
     <tr>
       <td><div align="center"><?php echo $row['物品编号'];?></div></td>
       <td><div align="center"><?php echo $row['物品名称'];?></div></td>
       <td><div align="center"><?php echo $row['颜色'];?></div></td>
       <td><div align="center"><?php echo $row['大小'];?></div></td>
       <!--<td><div align="center"><?php /*?><?php echo $row['产地'];?><?php */?></div></td>-->
       <td><div align="center"><?php echo $row['售价'];?></div></td>
       <td><div align="center"><?php echo $row['折扣'];?></div></td>
```

```
        <td><div align="center"><? php echo $row['数量']; ? ></div></td>
        <td><div align="center"><? php echo $row['店铺号']; ? ></div></td>
        <! --<td><div align="center"><? php /* ? ><? php echo $row['分类编号']; ? >
<? php */? ></div></td>-->
      </tr>
      </tbody>
    <? php
      $i= $i+1;
      if ( $i== $pagesize)
        break;
      }
    ? >
    </table>
    <div align="center">
    [第<? php echo $pageno; ? >页/共<? php echo $pagecount; ? >页]
    <? php
    $href= $PHP_SELF."? 物品编号 =".urlencode($物品编号);
    if ( $pageno<>1) {
    ? >
      <a href="<? php echo $href; ? >&pageno=1">首页</a>
      <a href="<? php echo $href; ? >&pageno=<? php echo $pageno-1; ? >">上一页</a>
    <? php
    }
    if ( $pageno<> $pagecount) {
    ? >
    <a href="<? php echo $href; ? >&pageno=<? php echo $pageno+1; ? >">下一页</a>
    <a href="<? php echo $href; ? >&pageno=<? php echo $pagecount; ? >">尾页</a>
    <? php
    }
    ? >
    [共找到<? php echo $rows; ? >个记录]
    </div>
    </body>
    </html>
```

代码运行效果见图 13.18。

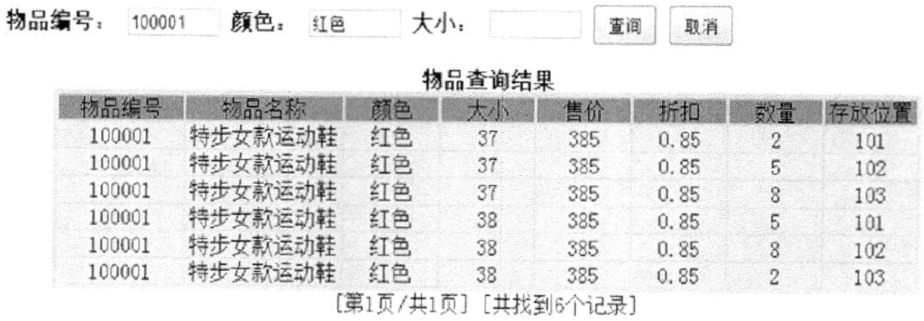

物品编号	物品名称	颜色	大小	售价	折扣	数量	存放位置
100001	特步女款运动鞋	红色	37	385	0.85	2	101
100001	特步女款运动鞋	红色	37	385	0.85	5	102
100001	特步女款运动鞋	红色	37	385	0.85	8	103
100001	特步女款运动鞋	红色	38	385	0.85	5	101
100001	特步女款运动鞋	红色	38	385	0.85	8	102
100001	特步女款运动鞋	红色	38	385	0.85	2	103

[第1页/共1页] [共找到6个记录]

图 13.18　商品查询

任务五　店内商品查询及销售

当销售员登录后，完成对自己店内库存商品的查询。如果销售员出售商品，则在本页面点击卖出按钮,进行销售。

查询页面,1_diannei_shangpin_xx.php 代码:

```html
<html>
<head>
<meta http-equiv="Content-Type" content="text/html; charset=gb2312">
<title>销售员查询结果</title>
<script type="text/JavaScript">
<!--
function MM_goToURL() { //v3.0
  var i, args=MM_goToURL.arguments; document.MM_returnValue = false;
  for (i=0; i<(args.length-1); i+=2)
eval(args[i]+".location='"+args[i+1]+"'");
}
//-->
</script>
</head>
<body>
<?php
//session_start();
?>
<?php
$销售员号=trim($销售员号);
  $link=mysql_connect("127.0.0.1","root","123456") or die("数据库服务器连接失败!<BR>");
  mysql_select_db("kcgl",$link) or die("数据库选择失败!<BR>");
  mysql_query("set names 'gbk'");
  $sql1="select 店铺号 from 销售员信息表 where 销售员号=$销售员号";
  $result=mysql_query($sql1,$link) or die("有错<br>");
  $row=mysql_fetch_array($result);
  $店铺号=$row['店铺号'];
  $sql="select 物品编号,物品名称,颜色,大小,售价,折扣,数量,店铺号 from 店铺物品表 where  店铺号='$店铺号'";
  $result=mysql_query($sql,$link) or die("有错<br>");
  $rows=mysql_num_rows($result);
  if ($rows==0)  {
    echo "没有满足条件的记录!";
    die();
  }
  $pagesize=12;   //每页的记录数(在此暂设为5,通常应设为10)
  $pagecount=ceil($rows/$pagesize);   //总页数
//$pageno 的值为当前页的页号
  if (!isset($pageno)||$pageno<1)
    $pageno=1;
  if ($pageno>$pagecount)
    $pageno=$pagecount;
  $offset=($pageno-1)*$pagesize;
  mysql_data_seek($result,$offset);
?>
<form name="form1" method="get">
  <div align="center"><strong>物品查询结果</strong> </div>
  <table style="table-layout:fixed;"  width="90%" border="3" cellpadding="0" cellspacing="0"  bordercolor="#99FFFF" align="center"  bgcolor="#CDE0F1">
    <thead bgcolor="#3399FF">
```

```
        <tr>
          <td width="80"><div align="center">物品编号</div></td>
          <td width="120"><div align="center">物品名称</div></td>
          <td width="60"><div align="center">颜色</div></td>
          <td width="60"><div align="center">大小</div></td>
          <!--<td width="90"><div align="center">产地</div></td>-->
          <td width="60"><div align="center">售价</div></td>
          <td width="60"><div align="center">折扣</div></td>
          <td width="60"><div align="center">数量</div></td>
          <td width="70"><div align="center">店铺号</div></td>
          <td width="60"><div align="center">操作</div></td>
          <!--<td width="90"><div align="center">分类编号</div></td>-->
        </tr>
      </thead>
      <?php
    $i=0;
    while($row=mysql_fetch_array($result)) {
    ?>
        <tr>
          <td><div align="center"><?php echo $row['物品编号']; ?></div></td>
          <td><div align="center"><?php echo $row['物品名称']; ?></div></td>
          <td><div align="center"><?php echo $row['颜色']; ?></div></td>
          <td><div align="center"><?php echo $row['大小']; ?></div></td>
          <!--<td><div align="center"><?php /*?><?php echo $row['产地']; ?><?php */?></div></td>-->
          <td><div align="center"><?php echo $row['售价']; ?></div></td>
          <td><div align="center"><?php echo $row['折扣']; ?></div></td>
          <td><div align="center"><?php echo $row['数量']; ?></div></td>
          <td><div align="center"><?php echo $row['店铺号']; ?></div></td>
          <!--<td><div align="center"><?php /*?><?php echo $row['分类编号']; ?><?php */?></div></td>-->
          <td><div align="center">
            <input name="submit" type="submit" onClick="MM_goToURL('parent','1_maichu.php?物品编号=<?php echo $row['物品编号']; ?>&物品名称=<?php echo $row['物品名称']; ?>&颜色=<?php echo $row['颜色']; ?>&大小=<?php echo $row['大小']; ?>&售价=<?php echo $row['售价']; ?>&折扣=<?php echo $row['折扣']; ?>&数量=<?php echo $row['数量']; ?>&店铺号=<?php echo $row['店铺号']; ?>');return document.MM_returnValue" value="卖出" />
          </div></td>
        </tr>
      </tbody>
    <?php
    $i=$i+1;
    if($i==$pagesize)
       break;
    }
    ?>
    </table>
    <p align="center">
  [第<?php echo $pageno; ?>页/共<?php echo $pagecount; ?>页]
<?php
//$href=$PHP_SELF."?物品编号=".urlencode($物品编号);
if($pageno<>1) {
```

```php
?>
<a href="<?php echo $href;?>&pageno=1">首页</a> <a href="<?php echo $href;?>&pageno=<?php echo $pageno-1;?>">上一页</a>
<?php
}
if($pageno<>$pagecount){
?>
<a href="<?php echo $href;?>&pageno=<?php echo $pageno+1;?>">下一页</a> <a href="<?php echo $href;?>&pageno=<?php echo $pagecount;?>">尾页</a>
<?php
}
?>
[共找到<?php echo $rows;?>个记录]</p>
</form>
<div align="center"></div>
<div align="center"></div>
</body>
</html>
```

查询结果，1_maichu.php 代码：

```php
<html xmlns="http://www.w3.org/1999/xhtml">
<head>
<meta http-equiv="Content-Type" content="text/html; charset=gb2312" />
<title>入库编辑</title>
</head>
<body>
<?php
//session_start();
?>
<?php
//$物品编号=$_SESSION['物品编号'];
//$物品名称=$_SESSION['物品名称'];
//$颜色=$_SESSION['颜色'];
//$大小=$_SESSION['大小'];
//$售价=$_SESSION['售价'];
//$折扣=$_SESSION['折扣'];
//$数量=$_SESSION['数量'];
//$店铺号=$_SESSION['店铺号'];
$物品编号=$_GET['物品编号'];
$物品名称=$_GET['物品名称'];
$颜色=$_GET['颜色'];
$大小=$_GET['大小'];
$售价=$_GET['售价'];
$折扣=$_GET['折扣'];
$数量=$_GET['数量'];
$店铺号=$_GET['店铺号'];
$link=mysql_connect("127.0.0.1","root","123456") or die("数据库服务器连接失败！<br>");
mysql_select_db("kcgl",$link) or die("数据库连接失败！<br>");
mysql_query("set names 'gbk'");
$sql="update 店铺物品表,销售表 set 店铺物品表.数量=店铺物品表.数量-1 where 店铺物品表.物品名称=销售表.物品名称 and 店铺物品表.物品编号=销售表.物品编号 and 店铺物品表.颜色=销售表.颜色
```

```
and 店铺物品表.大小=销售表.大小 and 店铺物品表.店铺号=销售表.店铺号 and 店铺物品表.物品名称='$物品名称' and 店铺物品表.物品编号='$物品编号' and 店铺物品表.颜色='$颜色' and 店铺物品表.大小='$大小' and 店铺物品表.店铺号='$店铺号'";
    if(mysql_query($sql,$link)){
    echo "更新成功！<br>";
    }
    else {
    echo "更新失败！";
    }
    $sql1="update 店铺物品表,销售表 set 销售表.数量=销售表.数量+1 where 店铺物品表.物品名称=销售表.物品名称 and 店铺物品表.物品编号=销售表.物品编号 and 店铺物品表.颜色=销售表.颜色 and 店铺物品表.大小=销售表.大小 and 店铺物品表.店铺号=销售表.店铺号 and 销售表.物品名称='$物品名称' and 销售表.物品编号='$物品编号' and 销售表.颜色='$颜色' and 销售表.大小='$大小' and 销售表.店铺号='$店铺号'";
    if(mysql_query($sql1,$link)){
    echo "更新成功！<br>";
    }
    else {
    echo "更新失败！<br>";
    }
    ?>
</body>
</html>
```

代码运行效果见图 13.19。

物品编号	物品名称	颜色	大小	售价	折扣	数量	店铺号	操作
100001	特步女款运动鞋	红色	37	385	0.85	2	101	卖出
100001	特步女款运动鞋	红色	38	385	0.85	5	101	卖出
100001	特步女款运动鞋	黄色	37	385	0.85	2	101	卖出
100001	特步女款运动鞋	黄色	38	385	0.85	4	101	卖出
100002	红蜻蜓男款皮鞋	黑色	40	370	0.00	7	101	卖出
100002	红蜻蜓男款皮鞋	黑色	41	370	0.00	5	101	卖出
100002	红蜻蜓男款皮鞋	棕色	41	370	0.00	7	101	卖出
100002	红蜻蜓男款皮鞋	棕色	42	370	0.00	5	101	卖出
100003	百丽女款休闲鞋	白色	37	240	0.00	4	101	卖出
100003	百丽女款休闲鞋	白色	38	240	0.00	7	101	卖出
100003	百丽女款休闲鞋	黑色	38	240	0.00	3	101	卖出
100003	百丽女款休闲鞋	黑色	39	240	0.00	1	101	卖出

图 13.19 店内商品查询

任务六 库 存 警 报

当库存商品数量太少时，系统给出警报，提醒管理者采取必要措施补货。

库存警报页面，1_kcjb.php 代码：

```
<!DOCTYPE html PUBLIC "-//W3C//DTD XHTML 1.0 Transitional//EN" "http://www.w3.org/TR/xhtml1/DTD/xhtml1-transitional.dtd">
<html xmlns="http://www.w3.org/1999/xhtml">
<head>
<meta http-equiv="Content-Type" content="text/html; charset=gb2312" />
```

```php
<title>无标题文档</title>
</head>
<body>
<? php
$销售员号=trim($销售员号);
$link=mysql_connect("127.0.0.1","root","123456") or die("数据库服务器连接失败！<BR>");
mysql_select_db("kcgl",$link) or die("数据库选择失败！<BR>");
mysql_query("set names 'gbk'");
$sql1="select 店铺号 from 销售员信息表 where 销售员号=$销售员号";
$result=mysql_query($sql1,$link) or die("有错<br>");
$row=mysql_fetch_array($result);
$店铺号=$row['店铺号'];
$sql="select 物品编号,物品名称,颜色,大小,数量 from 店铺物品表 where 数量<6 and 店铺号=$店铺号";
$result=mysql_query($sql,$link) or die("有错<br>");
$rows=mysql_num_rows($result);
if ($rows==0)  {
   echo "没有满足条件的记录!";
   die();
}
$pagesize=12；  //每页的记录数(在此暂设为5,通常应设为10)
$pagecount=ceil($rows/$pagesize);   //总页数
//$pageno 的值为当前页的页号
if (! isset($pageno)||$pageno<1)
   $pageno=1;
if ($pageno>$pagecount)
   $pageno=$pagecount;
$offset=($pageno-1)*$pagesize;
mysql_data_seek($result,$offset);
?>
<div align="center"><strong><font size="5">库存警报</font></strong> </div>
<table style="table-layout:fixed;"  width="90%" border="3" cellpadding="0" cellspacing="0" bordercolor="#99FFFF" align="center"  bgcolor="#D2D2D2">
<thead bgcolor="#CF6647">
  <tr>
    <td width="80"><div align="center"><font color="#CCFFFF" size="4">物品编号</font></div></td>
    <td width="90"><div align="center"><font color="#CCFFFF" size="4">物品名称</font></div></td>
    <td width="60"><div align="center"><font color="#CCFFFF" size="4">颜色</font></div></td>
    <td width="60"><div align="center"><font color="#CCFFFF" size="4">大小</font></div></td>
    <td width="60"><div align="center"><font color="#CCFFFF" size="4">数量</font></div></td>
    <td width="100"><div align="center"><font color="#CCFFFF" size="4">库存警报</font></div></td>
  </tr>
</thead>
<? php
  $i=0;
```

```
        while($row=mysql_fetch_array($result))  {
  ?>
    <tr>
      <td><div align="center"><?php echo $物品编号=$row['物品编号'];?></div></td>
      <td><div align="center"><?php echo $物品名称=$row['物品名称'];?></div></td>
      <td><div align="center"><?php echo $row['颜色'];?></div></td>
      <td><div align="center"><?php echo $row['大小'];?></div></td>
      <td><div align="center"><?php echo $row['数量'];?></div></td>
      <td><div align="center"><?php echo "存在";?></div></td>
    </tr>
    <?php
    $i=$i+1;
    if ($i==$pagesize)
      break;
    }
  ?>
</table>
<div align="center">
[第<?php echo $pageno;?>页/共<?php echo $pagecount;?>页]
<?php
$href=$PHP_SELF."?物品编号=".urlencode($物品编号);
if ($pageno<>1) {
?>
  <a href="<?php echo $href;?>&pageno=1">首页</a>
  <a href="<?php echo $href;?>&pageno=<?php echo $pageno-1;?>">上一页</a>
<?php
}
if ($pageno<>$pagecount) {
?>
<a href="<?php echo $href;?>&pageno=<?php echo $pageno+1;?>">下一页</a>
<a href="<?php echo $href;?>&pageno=<?php echo $pagecount;?>">尾页</a>
<?php
}
?>
[共找到<?php echo $rows;?>个记录]
</div>
</body>
</html>
```

代码运行效果见图 13.20。

库存警报

物品编号	物品名称	颜色	大小	数量	库存警报
100001	特步女款运动鞋	红色	37	2	存在
100001	特步女款运动鞋	红色	38	5	存在
100001	特步女款运动鞋	黄色	37	2	存在
100001	特步女款运动鞋	黄色	38	4	存在
100002	红蜻蜓男款皮鞋	黑色	41	5	存在
100002	红蜻蜓男款皮鞋	棕色	42	2	存在
100003	百丽女款休闲鞋	白色	37	4	存在
100003	百丽女款休闲鞋	黑色	38	3	存在
100003	百丽女款休闲鞋	黑色	39	1	存在

图 13.20　库存警报

任务七 修改商品价格

商品查询页面，2_update_jiage.html 代码：

```html
<html>
<head>
<title>修改价格</title>
</head>
<body>
<form action="2_update_jiage.php" target="right_2" method="get">
<font color="darkblue" size="2"><b>物品编号：</b></font>
<input type="text" name="物品编号" title="物品编号">
<input type="submit" value="查询">
<input type="reset" value="重置">
</form>
</body>
</html>
```

商品修改页面，2_update_jiage.php 代码：

```php
<html xmlns="http://www.w3.org/1999/xhtml">
<head>
<meta http-equiv="Content-Type" content="text/html; charset=gb2312" />
<title>信息查询</title>
</head>
<body>
<?php
session_start();
?>
<?php
$物品编号 = $_GET['物品编号'];
if($物品编号 == ""){
echo "物品编号不能为空！";
die();
}
$link = mysql_connect("127.0.0.1","root","123456") or die("数据库服务器连接失败！<br>");
mysql_select_db("kcgl", $link) or die("数据库选择成功！<br>");
mysql_query("set names 'gbk'");
$sql = "select 物品信息表.物品编号,物品名称,物品信息表.颜色,物品信息表.大小,售价,折扣 from 物品信息表 where 物品编号='$物品编号'";
$result = mysql_query($sql, $link) or die("有错<br>");
$rows = mysql_num_rows($result);
if($rows == 0){
echo "无满足条件的记录！";
die();
}
$pagesize = 10;
$pagecount = ceil($rows/$pagesize);
if(!isset($pageno) || $pageno<1)
$pageno = 1;
```

```php
    if($pageno>$pagecount)
    $pageno=$pagecount;
    $offset=($pageno-1)*$pagesize;
    mysql_data_seek($result,$offset);
?>
<form name="form1" action="2_update_jiage_1.php" method="get">
<div align="center";><strong>修改价格</strong></div>
<table style="table-layout:fixed;" width="90%" border="3" cellpadding="0" cellspacing="0" bordercolor="#99FFFF" align="center"  bgcolor="#CDE0F1">
<thead bgcolor="#3399FF">
<tr>
  <td><div align="center">物品编号</div></td>
  <td width="120"><div align="center">物品名称</div></td>
  <td><div align="center">颜色</div></td>
  <td><div align="center">大小</div></td>
  <td><div align="center">售价</div></td>
  <td><div align="center">折扣</div></td>
</tr>
</thead>
<?php
    while($row=mysql_fetch_array($result))
    {
?>
<tr>
  <td><div align="center"><input name="物品编号" type="text" value="<?php echo $row['物品编号'];?>" size="9" maxlength="9" /></div></td>
  <td><div align="center"><?php echo $row['物品名称'];?></div></td>
  <td><div align="center"><?php echo $row['颜色'];?></div></td>
  <td><div align="center"><?php echo $row['大小'];?></div></td>
  <td><div align="center"><input name="售价" type="text" value="<?php echo $row['售价'];?>" size="9" maxlength="9" /></div></td>
  <td><div align="center"><input name="折扣" type="text" value="<?php echo $row['折扣'];?>" size="9" maxlength="9" /></div></td>
<?php
    }
?>
</tr>
</table>
<br />
<div align="center">
<input name="wpbh0" type="hidden" value="<?php echo $物品编号;?>" />
<input name="sj0" type="hidden" value="<?php echo $售价;?>" />
<input name="zk0" type="hidden" value="<?php echo $折扣;?>" />
<input name="submit" type="submit" value="保存" />
<input name="reset" type="reset" value="取消" />
</div>
</form>
</body>
</html>
```

修改结果页面,2_update_jiage_1.php 代码:

```php
<html xmlns="http://www.w3.org/1999/xhtml">
<head>
<meta http-equiv="Content-Type" content="text/html; charset=gb2312" />
<title>编辑</title>
</head>
<body>
<? php
session_start();
? >
<? php
$物品编号 = $_GET['物品编号'];
$售价 = $_GET['售价'];
$折扣 = $_GET['折扣'];
if($物品编号=="" || $售价=="" || $折扣==""){
echo "物品编号及其售价、折扣均不能为空！<br>";
die();
}
$link = mysql_connect("127.0.0.1","root","123456") or die("数据库服务器连接失败！<br>");
mysql_select_db("kcgl", $link) or die("数据库连接失败！<br>");
mysql_query("set names 'gbk'");
$sql="update 物品信息表 set 售价='$售价',折扣='$折扣' where 物品编号='$物品编号'";
// $sql1="update 店铺物品表 set 售价='$售价',折扣='$折扣' where 物品编号='$物品编号'";
if(mysql_query($sql,$link)){
echo "价格修改成功！<br>";
}
else {
echo "价格修改失败！<br>";
}
$sql="update 店铺物品表 set 售价='$售价',折扣='$折扣' where 物品编号='$物品编号'";
if(mysql_query($sql,$link)){
echo "价格修改成功！<br>";
}
else {
echo "价格修改失败！<br>";
}
? >
</body>
</html>
```

代码运行效果见图 13.21。

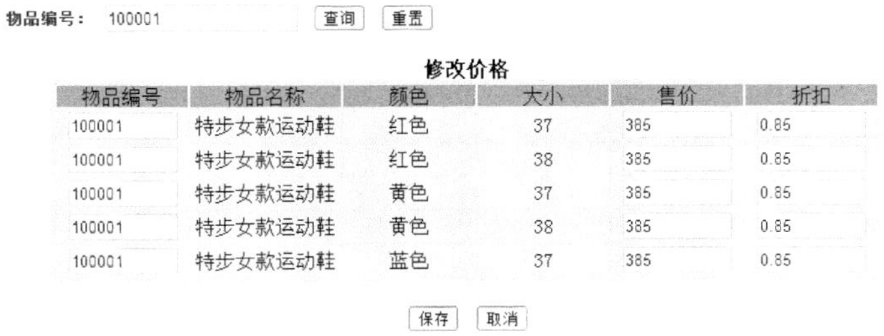

图 13.21　修改价格

任务八　销售情况饼状图

对销售情况进行统计分析，并绘制成饼状图，让管理者可以一目了然地看到销售情况。本页面需要使用"libchart"。下载地址：http://download.csdn.net/download/hdznz/3563380。

销售情况表，PieChartTest.php 代码：

```php
<?php
    include "../libchart/classes/libchart.php";
$link=mysql_connect("127.0.0.1","root","123456");
mysql_select_db("kcgl",$link);
mysql_query("set names 'gbk'");
$sql="SELECT    店铺号,数量 ,sum(数量) as 总数
FROM
'销售表'
GROUP BY
'销售表'.'店铺号'";
$result=mysql_query($sql,$link);
    $chart = new PieChart();
    $dataSet = new XYDataSet();
    while($row=mysql_fetch_array($result))
  {
    $dataSet->addPoint(new Point( $row['店铺号'] , $row['总数']));
    $chart->setDataSet($dataSet);
    $chart->setTitle("dian pu xiao shou qing kuang bing zhuang tu");
    $chart->render("generated/demo3.jpg");}
?><head>
    <title><center>dian pu xiao shou qing kuang bing zhuang tu</center></title>
<meta http-equiv="Content-Type" content="text/html;charset=gb2312" />
</head>
<body>
    <img alt="Pie chart"  src="generated/demo3.jpg" style="border: 1px solid gray;"/>
</body>
</html>
```

代码运行效果见图 13.22。

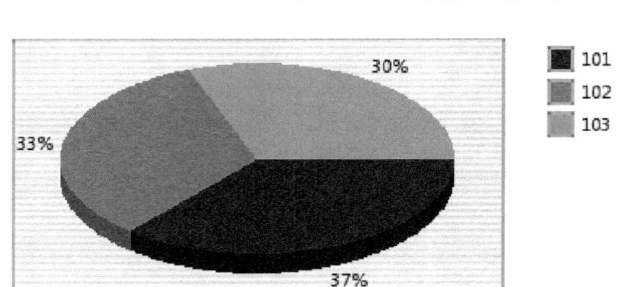

图 13.22　销售情况

任务九　新闻管理

主要完成系统内部新闻的发布、修改、删除等功能。

新闻编辑页面，sample01.php 代码：

```php
<?php
/*
 * FCKeditor - The text editor for Internet - http://www.fckeditor.net
 * Copyright (C) 2003-2010 Frederico Caldeira Knabben
 *
 * == BEGIN LICENSE ==
 *
 * Licensed under the terms of any of the following licenses at your
 * choice:
 *
 *  - GNU General Public License Version 2 or later (the "GPL")
 *    http://www.gnu.org/licenses/gpl.html
 *
 *  - GNU Lesser General Public License Version 2.1 or later (the "LGPL")
 *    http://www.gnu.org/licenses/lgpl.html
 *
 *  - Mozilla Public License Version 1.1 or later (the "MPL")
 *    http://www.mozilla.org/MPL/MPL-1.1.html
 *
 * == END LICENSE ==
 *
 * Sample page.
 */
include("../../fckeditor.php") ;
?>
<!DOCTYPE HTML PUBLIC "-//W3C//DTD HTML 4.0 Transitional//EN">
<html>
    <head>
        <title>FCKeditor - Sample</title>
        <meta http-equiv="Content-Type" content="text/html; charset=utf-8">
        <meta name="robots" content="noindex, nofollow">
        <link href="../sample.css" rel="stylesheet" type="text/css" />
    </head>
    <body>
        <h1>上传新闻</h1>
        This sample displays a normal HTML form with an FCKeditor with full features enabled.
        <hr>
        <form action="sampleposteddata.php" method="post" target="right_2">
        公告ID：<input name="公告ID" type="text" size="20" maxlength="20">
        标题：<input name="标题" type="text" size="20" maxlength="20">
        上传人：<input name="上传人" type="text" size="20" maxlength="20">
<?php
// Automatically calculates the editor base path based on the _samples directory.
```

```php
// This is usefull only for these samples. A real application should use something like this:
// $oFCKeditor->BasePath = '/fckeditor/' ;    // '/fckeditor/' is the default value.

$sBasePath = $_SERVER['PHP_SELF'] ;
$sBasePath = substr( $sBasePath, 0, strpos( $sBasePath, "_samples" ) ) ; //截取,位置取得
$oFCKeditor = new FCKeditor('FCKeditor1') ;
$oFCKeditor->Width = '90%';
$oFCKeditor->Height = '300px';
$oFCKeditor->BasePath = $sBasePath ;
$oFCKeditor->Value = '<p>This is some <strong>sample text</strong>. You are using <a href="http://www.fckeditor.net/">FCKeditor</a>.</p>' ;
$oFCKeditor->Create() ;
?>
                    <br>
                    <input type="submit" value="Submit">
        </form>
    </body>
</html>
```

新闻处理页面，neweditor.php 代码：

```html
<!DOCTYPE html PUBLIC "-//W3C//DTD XHTML 1.0 Transitional//EN"
"http://www.w3.org/TR/xhtml1/DTD/xhtml1-transitional.dtd">
<html xmlns="http://www.w3.org/1999/xhtml">
<head>
<meta http-equiv="Content-Type" content="text/html; charset=gb2312" />
<title>无标题文档</title>
</head>
<body vlink="lightblack" link="#666666">
<style type="text/css">a:link,a:visited{ text-decoration:none; font-family:微软雅黑;}
a:hover{ text-decoration:underline;   /*鼠标放上去有下划线*/}
<!--html{font-size:62.5%}
body{font-size:1.2em; color:#294f88;}
.sample{margin:30px; border:0px solid #92cdec; background:; padding:30px 30px 0 30px; width:500px; height:600px;}
    h1{margin:0;padding:0 0 30px 0;font-size:2em}
    h2{margin:0;padding:10px 0 5px 0;font-size:1.5em;clear:both}
    p{font-size:1.2em;line-height:170%;margin-bottom:30px;clear:both}
    ul,ol{padding:0;margin:0;font-size:1.2em}

    .sample ul,.sample ol{list-style-type:none; background:url(img/dotted.gif) 0 0 repeat-x; width:400px; float:left;margin:0 10px 30px 0}
    .sample li{background:url(img/dotted.gif) bottom left repeat-x;width:400px;float:left;padding:0}
    .sample li a{padding:8px 20px 8px 0;width:400px;float:left;background:url(img/news.gif) right 12px no-repeat;text-decoration:none;color:#294f88}
    .sample li a:hover{background-image:url(img/ys1.gif)}-->
</style>
<div style="position:relative; top:2px; width:980px; left:0px; height:340px;">
```

```php
<table width="750">
<tr>
  <td align="center">公告栏</td>
</tr>
</table>
    <? php
$link=mysql_connect("127.0.0.1","root","123456") or die("数据库服务器连接失败！<BR>");
mysql_select_db("kcgl",$link) or die("数据库选择失败！<BR>");
mysql_query("set names 'gbk'");
$sql="select * from 公告表 ";
$result=mysql_query($sql,$link);
$row = mysql_fetch_array($result);
mysql_data_seek($result,0);
while($row = mysql_fetch_array($result))
{
?>
<!--<ul class="loremipsum" style="height:1px;">
<li><li>
</ul>-->
<table width="80%" border="0"  cellspacing="0" cellpadding="0" class="red" style="padding-left:60px">
          <tbody>
            <tr>
              <td width="6%" height="10" align="center"><img src="img/ys1.gif" width="10" height="10" border="0"></td>
              <td width="94%" height="10" align="left"><table width="94%" class="red" cellpadding="0" cellspacing="4" border="0">
                <tbody>
                  <tr>
                    <td width="729" height="10" align="left"><a href="text.php?标题=<? php echo $row['标题']; ?>" target="right_2" style="text-decoration:none"><? php echo $row['标题']; ?></a></td>
                    <td width="39" align="left"><div style="white-space:nowrap"><img src=" img/news.gif" border="0"></div></td>
                    <td width="29" align="left"><div style="white-space:nowrap"><? php echo $row['上传时间']; ?></div></td>
                  </tr></tbody></table></td>
            </tr>
            <tr><td colspan="2" style="background:url(img/dotted.gif) repeat-x" height="2"></td>
  </tbody></table>
  <br />
   <!--<table width="740" border="0">
    <tr>
      <td><font size="2"> <a target="right_2" href="text.php?标题=<? php echo $row['标题']; ?>"><? php echo $row['标题']; ?></a></font>    </td>
      <td align="right"><font size="2" color="orange"><? php echo $row['上传时间']; ?></font></td>
    </tr>
  </table>-->
```

```html
<!--<a href="text.php?name=<?php echo $row['公告ID'];?>" target="_self"><table width="200px" height="100px"><tr><td><?php echo $row['标题'];?></td><td><?php echo $row['上传时间'];?></td></tr></table></a><br/>-->
        <?php
        }
        ?>
    </div>
    <!--<br/><br/><div style="margin-bottom:0; font-size:1px;" align="right"><a href="">更多>></a>  </div>-->
</body>
</html>
```

新闻浏览页面，text.php 代码：

```html
<!DOCTYPE html PUBLIC "-//W3C//DTD XHTML 1.0 Transitional//EN" "http://www.w3.org/TR/xhtml1/DTD/xhtml1-transitional.dtd">
<html xmlns="http://www.w3.org/1999/xhtml">
<head>
<meta http-equiv="Content-Type" content="text/html; charset=gb2312" />
<title>无标题文档</title>
</head>
<body>
<?php
$name = $_GET['标题'];
$link = mysql_connect("127.0.0.1","root","123456")
    or die("数据库服务器连接失败！<BR>");
mysql_select_db("kcgl", $link) or die("数据库选择失败！<BR>");
mysql_query("set names 'gbk'");
$sql = "select * from 公告表 where 标题='$name'";
$result = mysql_query($sql, $link);
$row = mysql_fetch_array($result);
?>
<table width="750" height="94" border="0">
    <tr>
        <td height="27" colspan="2"><div align="center">
        <font size="5"><?php echo $row['标题'];?></font>
        </div></td>
    </tr>
    <tr>
        <td height="24" style="padding-left:180px"><font size="3">上传人：
            <?php echo $row['上传人'];?></font></td>
        <td><font size="3">上传时间：
            <?php echo $row['上传时间'];?></font></td>
    </tr>
    <tr>
        <td height="35" colspan="2"><?php echo $row['正文'];?></td>
    </tr>
</table>
</body>
</html>
```

代码运行效果见图 13.23。

图 13.23 新闻管理

公告修改页面,samplesposteddata.php 代码:

```php
<? php
/*
 * FCKeditor - The text editor for Internet - http://www.fckeditor.net
 * Copyright (C) 2003-2010 Frederico Caldeira Knabben
 *
 * == BEGIN LICENSE ==
 *
 * Licensed under the terms of any of the following licenses at your
 * choice:
 *
 *  - GNU General Public License Version 2 or later (the "GPL")
 *    http://www.gnu.org/licenses/gpl.html
 *
 *  - GNU Lesser General Public License Version 2.1 or later (the "LGPL")
 *    http://www.gnu.org/licenses/lgpl.html
 *
 *  - Mozilla Public License Version 1.1 or later (the "MPL")
 *    http://www.mozilla.org/MPL/MPL-1.1.html
 *
 * == END LICENSE ==
 *
 * This page lists the data posted by a form.
 */
?>
<!DOCTYPE HTML PUBLIC "-//W3C//DTD HTML 4.0 Transitional//EN">
<html>
    <head>
        <title>FCKeditor - Samples - Posted Data</title>
        <meta http-equiv="Content-Type" content="text/html; charset=utf-8">
        <meta name="robots" content="noindex, nofollow">
        <link href="../sample.css" rel="stylesheet" type="text/css">
    </head>
    <body>
        <h1>FCKeditor - Samples - Posted Data</h1>
        This page lists all data posted by the form.
        <hr>
        <table border="1" cellspacing="0" id="outputSample">
            <colgroup><col width="80"><col></colgroup>
```

```html
                    <thead>
                        <tr>
                            <th>Field Name</th>
                            <th>Value</th>
                        </tr>
                    </thead>
```
```php
<?php

if( isset( $_POST ) )
    $postArray = &$_POST ;            // 4.1.0 or later, use $_POST
else
    $postArray = &$HTTP_POST_VARS ;   // prior to 4.1.0, use HTTP_POST_VARS

foreach( $postArray as $sForm => $value )
{
    if( get_magic_quotes_gpc() )//当 magic_quotes_gpc 打开时,所有的 '(单引号),"(双引号),\
(反斜线) and 空字符会自动转为含有反斜线的转义字符。
        $postedValue = htmlspecialchars( stripslashes( $value ) ) ;//函数删除由 addslashes()
函数添加的反斜杠。htmlspecialchars 把一些预定义的字符转换为 HTML 实体。
    else
        $postedValue = addslashes( $value ) ;//在指定的预定义字符前添加反斜杠
?>
                        <tr>
                            <th><?php echo htmlspecialchars( $sForm) ?></th>
                            <td><pre><?php echo $postedValue? ></pre></td>
                        </tr>
<?php
}
$公告ID = $_POST['公告ID'];
$标题 = $_POST['标题'];
$上传人 = $_POST['上传人'];
$link = mysql_connect("127.0.0.1","root","123456")
    or die("数据库服务器连接失败!<BR>");
mysql_select_db("kcgl", $link) or die("数据库选择失败!<BR>");
mysql_query("set names 'utf8'");
$sql = "INSERT INTO '公告表' VALUES ('$公告ID','$标题','$postedValue','$上传人',current_date())";
if (mysql_query($sql, $link))
    echo "增加成功!";
else
    echo '增加失败!';
?>
                </table>
            </body>
</html>
```

公告处理页面,sampleposteddata1.php 代码:

```php
<?php
/*
 * FCKeditor - The text editor for Internet - http://www.fckeditor.net
```

```
 * Copyright (C) 2003-2010 Frederico Caldeira Knabben
 *
 * == BEGIN LICENSE ==
 *
 * Licensed under the terms of any of the following licenses at your
 * choice:
 *
 *  - GNU General Public License Version 2 or later (the "GPL")
 *    http://www.gnu.org/licenses/gpl.html
 *  - GNU Lesser General Public License Version 2.1 or later (the "LGPL")
 *    http://www.gnu.org/licenses/lgpl.html
 *  - Mozilla Public License Version 1.1 or later (the "MPL")
 *    http://www.mozilla.org/MPL/MPL-1.1.html
 *
 * == END LICENSE ==
 *
 * This page lists the data posted by a form.
 */
?>
<!DOCTYPE HTML PUBLIC "-//W3C//DTD HTML 4.0 Transitional//EN">
<html>
    <head>
        <title>FCKeditor - Samples - Posted Data</title>
        <meta http-equiv="Content-Type" content="text/html; charset=gb2312">
        <meta name="robots" content="noindex, nofollow">
        <link href="../sample.css" rel="stylesheet" type="text/css">
    </head>
    <body>
        <h1>FCKeditor - Samples - Posted Data</h1>
        This page lists all data posted by the form.
        <hr>
        <table border="1" cellspacing="0" id="outputSample">
            <colgroup><col width="80"><col></colgroup>
            <thead>
                <tr>
                    <th>Field Name</th>
                    <th>Value</th>
                </tr>
            </thead>
<?php
if ( isset( $_POST ) )
    $postArray = &$_POST ;          // 4.1.0 or later, use $_POST
else
    $postArray = &$HTTP_POST_VARS ; // prior to 4.1.0, use HTTP_POST_VARS

foreach ( $postArray as $sForm => $value )
{
    if ( get_magic_quotes_gpc() )
        $postedValue = htmlspecialchars( stripslashes( $value ) ) ;
    else
        $postedValue = addslashes( $value ) ;
```

```php
?>
                        <tr>
                            <th><?php echo htmlspecialchars($sForm)?></th>
                            <td><pre><?php echo $postedValue?></pre></td>
                        </tr>
<?php
}
$公告ID=$_POST['公告ID'];
$标题= $_POST['标题'];
$上传人= $_POST['上传人'];
$link=mysql_connect("127.0.0.1","root","123456")
    or die("数据库服务器连接失败！<BR>");
  mysql_select_db("kcgl",$link) or die("数据库选择失败！<BR>");
  mysql_query("set names 'gb2312'");
/* $sql="INSERT INTO 'news' VALUES ('$id','$title','$postedValue','$date','$author')"; */
  $sql="update 公告表 set 公告ID='$公告ID',标题='$标题',正文='$postedValue',上传人='$上传人',上传时间=current_date() where 公告ID='$公告ID'";
  if (mysql_query($sql,$link))
    echo "update success!";
  else
    echo 'update failed! ';
?>
    </body>
</html>
```

删除页面，delete_news.php 代码：

```php
<?php
$标题= $_GET['标题'];
$link=mysql_connect("127.0.0.1","root","123456")
    or die("数据库服务器连接失败！<BR>");
  mysql_select_db("kcgl",$link) or die("数据库选择失败！<BR>");
  mysql_query("set names 'gb2312'");
  $sql="delete from 公告表 where 标题='$标题'";
  if (mysql_query($sql,$link))
    echo "删除成功！";
  else
    echo '删除失败！ ';
?>
```

代码运行效果见图 13.24。

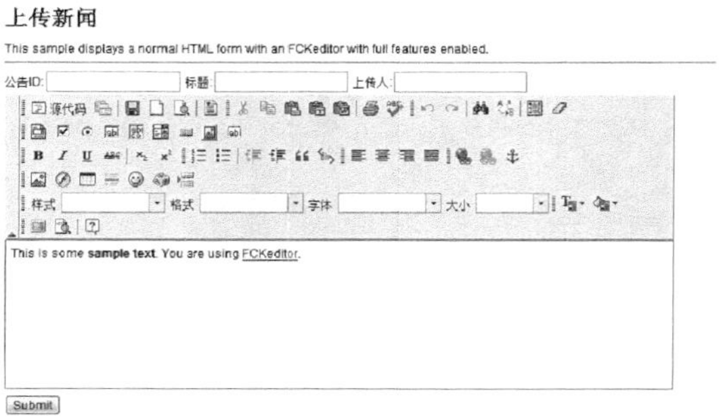

图 13.24　上传新闻

任务十 入库管理

主要完成商品入库信息的记录和管理。

入库页面,ruku.php 代码:

```php
<!DOCTYPE html PUBLIC "-//W3C//DTD XHTML 1.0 Transitional//EN" "http://www.w3.org/TR/xhtml1/DTD/xhtml1-transitional.dtd">
<html xmlns="http://www.w3.org/1999/xhtml">
<head>
<meta http-equiv="Content-Type" content="text/html; charset=gb2312" />
<title>无标题文档</title>
</head>

<body>
<?php
$link=mysql_connect("127.0.0.1","root","123456") or die("数据库服务器连接失败!<br>");
mysql_select_db("kcgl",$link) or die("数据库选择成功!<br>");
  mysql_query("set names 'gbk'");
  $sql="select * from 入库表";
  $result=mysql_query($sql,$link) or die("有错<br>");
$rows=mysql_num_rows($result);
  if ($rows==0)  {
    echo "没有满足条件的记录!";
    die();
  }
  $pagesize=10;   //每页的记录数(在此暂设为5,通常应设为10)
  $pagecount=ceil($rows/$pagesize);   //总页数
  //$pageno 的值为当前页的页号
  if (!isset($pageno)||$pageno<1)
    $pageno=1;
  if ($pageno>$pagecount)
    $pageno=$pagecount;
  $offset=($pageno-1)*$pagesize;
mysql_data_seek($result,$offset);

?>

<style type="text/css">
.tableStyle
{
border-collapse:collapse;
width:700px;
}
td
{
width:60px;
font-size:12px;
height:25px;
```

```
border:1px solid #CCD5E8;
}
.btn {
font-size:12pt; color: #003399;
border: 1px #003399 solid;
color: #006699;
border-bottom: #93bee2 1px solid;
border-left: #93bee2 1px solid;
border-right: #93bee2 1px solid;
border-top: #93bee2 1px solid;
background-color: #e8f4ff;
cursor: pointer;
font-style: normal ;
width:20px;
height:22px;
font-family:Verdana;font-family:Georgia;_font-family:Tahoma;
padding:0 10px 1px;padding:3px 3px 1px;_padding:0 4px 1px;
line-height:18px;line-height:14px;_line-height:16px;
}
</style>
</head>
<body>
<form name="form1" action="ruku_a.php" method="get">
<div align="center";><strong>物品入库</strong></div>
<table class="tableStyle" id="OwnershipStructure">
<tr>
<td align="center" width="0" rowspan="2" id="StructureLeft1" style="border:hidden"> </td>
<td align="center" width="23" rowspan="2" id="StructureLeft2"> </td>
<td align="center" width="60">入库编号 </td>
<td align="center" width="70">物品编号 </td>
<td align="center" width="100">物品名称 </td>
<td align="center" width="60">颜色</td>
<td align="center" width="60">大小 </td>
<td align="center" width="60">进价 </td>
<td align="center" width="60">货位编号 </td>
<td align="center" width="100">入库日期 </td>
<td align="center" width="60">入库数量 </td>
<td align="center" width="107">负责人 </td>
</tr>
  <tr id="StructureRight">
<td>
<input  name="入库编号[]"   style="width:60px;height:20px;" id="Text1" type="text" /></td>
<td>
<input  name="物品编号[]" style="width:70px;height:20px;" id="Text2" type="text" /></td>
<td>
<input name="物品名称[]" style="width:100px;height:20px;" id="Text3" type="text" /></td>
<td>
<input name="颜色[]" style="width:60px;height:20px;" id="Text4" type="text" /></td>
<td>
<input name="大小[]"   style="width:60px;height:20px;" id="Text5" type="text" /></td>
```

```html
<td>
<input name="进价[]" style="width:60px;height:20px;" id="Text6" type="text" /></td>
<td>
<input name="货位编号[]" style="width:60px;height:20px;" id="Text7" type="text" /></td>
<td>
<input name="入库日期[]" style="width:100px;height:20px;" id="Text8" type="text" /></td>
<td>
<input name="入库数量[]" style="width:60px;height:20px;" id="Text9" type="text" /></td>
<td>
<input name="负责人[]" style="width:60px;height:20px;" id="Text10" type="text" /></td>
<td><input id="btnAddRow" class="btn"
onclick="AddStructureRow()" type="button" value="+" />
</td>
</tr>
<!--<tr>
<td>评估机构</td><td></td><td></td><td></td>
</tr>
<tr><td>评估机构</td><td></td><td></td><td></td>
</tr> -->
</table>
<script language="javascript" type="text/javascript">
//表单操作
function AddStructureRow()
{
var obj=document.getElementById("OwnershipStructure");
var tr= obj.rows["StructureRight"];
//alert(tr.rowIndex);
var count=document.getElementById("StructureLeft1").getAttribute("rowspan");
document.getElementById("StructureLeft1").setAttribute("rowSpan",parseInt(count)+1);
document.getElementById("StructureLeft2").setAttribute("rowSpan",parseInt(count)+1);
//插入行 code form www.jb51.net
var tr = obj.insertRow(tr.rowIndex+1);
var trId = "trStructure" + tr.rowIndex;
tr.setAttribute("id",trId);
var td0 = tr.insertCell(0);
td0.setAttribute("align","left");
//td0.setAttribute("colSpan","4");
td0.innerHTML = "<input ID='txtName' name='入库编号[]' style='width:60px;height:20px;' type='text'/> ";
var td0 = tr.insertCell(1);
td0.setAttribute("align","left");
//td0.setAttribute("colSpan","4");
td0.innerHTML = "<input ID='txtName' name='物品编号[]' style='width:70px;height:20px;' type='text'/> ";
var td0 = tr.insertCell(2);
td0.setAttribute("align","left");
//td0.setAttribute("colSpan","4");
td0.innerHTML = "<input ID='txtName' name='物品名称[]' style='width:100px;height:20px;' type='text'/> ";
var td0 = tr.insertCell(3);
td0.setAttribute("align","left");
```

```
        //td0.setAttribute("colSpan","4");
        td0.innerHTML = "<input ID='txtName' name='颜色[]' style='width:60px;height:20px;' type='text'/> ";
        var td0 = tr.insertCell(4);
        td0.setAttribute("align","left");
        //td0.setAttribute("colSpan","4");
        td0.innerHTML = "<input ID='txtName'  name='大小[]'  style='width:60px;height:20px;' type='text'/> ";
        var td0 = tr.insertCell(5);
        td0.setAttribute("align","left");
        //td0.setAttribute("colSpan","4");
        td0.innerHTML = "<input ID='txtName' name='进价[]' style='width:60px;height:20px;' type='text'/> ";
        var td0 = tr.insertCell(6);
        td0.setAttribute("align","left");
        //td0.setAttribute("colSpan","4");
        td0.innerHTML = "<input ID='txtName' name='货位编号[]' style='width:60px;height:20px;' type='text'/> ";
        var td0 = tr.insertCell(7);
        td0.setAttribute("align","left");
        //td0.setAttribute("colSpan","4");
        td0.innerHTML = "<input ID='txtName' name='入库日期[]' style='width:100px;height:20px;' type='text'/> ";
        var td0 = tr.insertCell(8);
        td0.setAttribute("align","left");
        //td0.setAttribute("colSpan","4");
        td0.innerHTML = "<input ID='txtName' name='入库数量[]' style='width:60px;height:20px;' type='text'/> ";
        var td1 = tr.insertCell(9);
        td1.setAttribute("align","left");
        //td1.setAttribute("colSpan","3");
        td1.innerHTML = "<input ID='txtName' name='负责人[]' style='width:60px;height:20px;' type='text'/>";
        var td1 = tr.insertCell(10);
        td1.setAttribute("align","left");
        td1.innerHTML = "<input id='btnDelRow' class='btn' type='button' value='-' onclick='DelStructureRow("+tr.rowIndex+")'/>";
    }
    function DelStructureRow(rowIndex)
    {
        var obj=document.getElementById("OwnershipStructure");
        obj.deleteRow(rowIndex);
        var count=document.getElementById("StructureLeft1").getAttribute("rowspan");
        document.getElementById("StructureLeft1").setAttribute("rowSpan",parseInt(count)-1);//
rowSpan 不要写成 rowspan,因为在 IE6 与 IE7 下会有问题
        document.getElementById("StructureLeft2").setAttribute("rowSpan",parseInt(count)-1);
    }
</script>
<div align="center">
<input name="submit" type="submit" value="保存" />
<input name="reset" type="reset" value="取消" />
```

```
</div>
</form>
</body>
</html>
```

入库编辑页面,ruku_a.php 代码:

```
<html xmlns="http://www.w3.org/1999/xhtml">
<head>
<meta http-equiv="Content-Type" content="text/html; charset=gb2312" />
<title>入库编辑</title>
</head>
<body>
<?php
session_start();
?>
<?php
$入库编号 = $_GET['入库编号'];
$物品编号 = $_GET['物品编号'];
$物品名称 = $_GET['物品名称'];
$颜色 = $_GET['颜色'];
$大小 = $_GET['大小'];
$进价 = $_GET['进价'];
$货位编号 = $_GET['货位编号'];
$入库日期 = $_GET['入库日期'];
$入库数量 = $_GET['入库数量'];
$负责人 = $_GET['负责人'];
if($入库编号=="" || $物品编号=="" || $物品名称=="" || $颜色=="" || $大小=="" || $进价=="" || $货位编号=="" || $入库日期=="" || $入库数量=="" || $负责人==""){
echo "入库编号及其物品编号、物品名称,颜色,大小,进价,货位编号,入库日期,入库数量,负责人均不能为空!<br>";
die();
}
$link=mysql_connect("127.0.0.1","root","123456") or die("数据库服务器连接失败!<br>");
mysql_select_db("kcgl",$link) or die("数据库连接失败!<br>");
mysql_query("set names 'gbk'");
  for($i=0;$i<count($_GET['入库编号']);$i++)
{$入库编号 = $_GET['入库编号'][$i];
$物品编号 = $_GET['物品编号'][$i];
$物品名称 = $_GET['物品名称'][$i];
$颜色 = $_GET['颜色'][$i];
$大小 = $_GET['大小'][$i];
$进价 = $_GET['进价'][$i];
$货位编号 = $_GET['货位编号'][$i];
$入库日期 = $_GET['入库日期'][$i];
$入库数量 = $_GET['入库数量'][$i];
$负责人 = $_GET['负责人'][$i];
@$value.="('$入库编号','$物品编号','$物品名称','$颜色','$大小','$进价','$货位编号','$入库日期','$入库数量','$负责人'),";
}
$sql="insert into 入库表(入库编号,物品编号,物品名称,颜色,大小,进价,货位编号,入库日期,入库数
```

```
量,负责人) values".rtrim($value,",");
    if(mysql_query($sql,$link)){
    echo "物品入库成功! <br>";
    }
    else {
    echo "物品入库失败! <br>";
    }
    for($i=0;$i<count($_GET['入库编号']);$i++)
    {$入库编号=$_GET['入库编号'][$i];
    $物品编号=$_GET['物品编号'][$i];
    $颜色=$_GET['颜色'][$i];
    $大小=$_GET['大小'][$i];
    @$value.="('$入库编号','$物品编号','$物品名称','$颜色','$大小','$进价','$货位编号','$入库日期','$入库数量','$负责人'),";
    $sql1="update 实际库存表,入库表 set 物品数量=$入库数量+物品数量 where 实际库存表.物品编号=入库表.物品编号 and 实际库存表.颜色=入库表.颜色 and 实际库存表.大小=入库表.大小";
    }
    if(mysql_query($sql1,$link)){
    echo "更新成功! <br>";
    }
    else {
    echo "更新失败! <br>";
    }
    ?>
    </body>
    </html>
```

代码运行效果见图 13.25。

图 13.25 物品入库

入库单查询页面,rukudan_xx.html 代码:

```
<html>
<head>
<title>入库单信息</title>
</head>
<body>
<form action="rukudan_xx.php" target="right_2" method="get">
<b>按入库编号查询:</b>
<input type="text" name="入库编号">
<input type="submit" value="查询">
<input type="reset" value="取消">
</form>
</body>
</html>
```

入库单查询结果页面, rukudan_xx.php 代码:

```php
<html>
<head>
<meta http-equiv="Content-Type" content="text/html; charset=gb2312">
<title>销售员查询结果</title>
</head>
<body>
<form action="3_dayin_rukudan.php" method="get">
<INPUT TYPE="button" id="print" value="导出 Excel" name="print" onClick="javascript:window.open('3_dayin_rukudan.php')" >
</form>
<? php
session_start();
//require("a.php");
//if(! isset($_SESSION['销售员号'])){
//echo"身份不符合,请重新选择身份!";
//header("refresh:2;url=a.htm");
//                     exit;
//}
    $入库编号 = $_GET['入库编号'];
if($入库编号 == ""){
echo "入库编号不能为空!";
die();
}
    $link=mysql_connect("127.0.0.1","root","123456") or die("数据库服务器连接失败!<br>");
mysql_select_db("kcgl",$link) or die("数据库选择成功!<br>");
    mysql_query("set names 'gbk'");
    $sql="select * from 入库表 where 入库表.入库编号='$入库编号' order by '$入库编号'";
    $result=mysql_query($sql,$link) or die("有错<br>");
 $rows=mysql_num_rows($result);
    if ($rows==0)  {
      echo "没有满足条件的记录!";
      die();
    }
    $pagesize=10;  //每页的记录数(在此暂设为5,通常应设为10)
    $pagecount=ceil($rows/$pagesize);   //总页数
//$pageno 的值为当前页的页号
   if (! isset($pageno)||$pageno<1)
      $pageno=1;
   if ($pageno>$pagecount)
      $pageno=$pagecount;
      $offset=($pageno-1)*$pagesize;
  mysql_data_seek($result,$offset);
 ?>
 <div align="center"><strong>入库单查询结果</strong> </div>
 <table style="table-layout:fixed;"   width="90%" border="3" cellpadding="0" cellspacing="0"   bordercolor="#99FFFF" align="center"   bgcolor="#CDE0F1">
 <thead bgcolor="#3399FF">
    <tr>
      <td width="80"><div align="center">入库编号</div></td>
```

```
        <td width="80"><div align="center">物品编号</div></td>
        <td width="120"><div align="center">物品名称</div></td>
        <td width="60"><div align="center">颜色</div></td>
        <td width="60"><div align="center">大小</div></td>
            <td width="60"><div align="center">进价</div></td>
            <td width="80"><div align="center">货位编号</div></td>
            <td width="90"><div align="center">入库日期</div></td>
            <td width="60"><div align="center">入库数量</div></td>
            <td width="70"><div align="center">负责人</div></td>
    </tr>
    </thead>
<?php
    $i=0;
    while($row=mysql_fetch_array($result))  {
?>
    <tr>
        <td><div align="center"><?php echo $row['入库编号'];?></div></td>
            <td><div align="center"><?php echo $row['物品编号'];?></div></td>
        <td><div align="center"><?php echo $row['物品名称'];?></div></td>
        <td><div align="center"><?php echo $row['颜色'];?></div></td>
        <td><div align="center"><?php echo $row['大小'];?></div></td>
        <td><div align="center"><?php echo $row['进价'];?></div></td>
        <td><div align="center"><?php echo $row['货位编号'];?></div></td>
        <td><div align="center"><?php echo $row['入库日期'];?></div></td>
        <td><div align="center"><?php echo $row['入库数量'];?></div></td>
        <td><div align="center"><?php echo $row['负责人'];?></div></td>
      </tr>
    </tbody>
<?php
    $i=$i+1;
    if ($i==$pagesize)
        break;
    }
?>
</table>
<?php
$_SESSION['入库编号']=mysql_result($result,0,"入库编号");
?>
<div align="center">
[第<?php echo $pageno;?>页/共<?php echo $pagecount;?>页]
<?php
$href=$PHP_SELF."?入库编号=".urlencode($入库编号);
if ($pageno<>1) {
?>
    <a href="<?php echo $href;?>&pageno=1">首页</a>
    <a href="<?php echo $href;?>&pageno=<?php echo $pageno-1;?>">上一页</a>
<?php
}
if ($pageno<>$pagecount) {
?>
<a href="<?php echo $href;?>&pageno=<?php echo $pageno+1;?>">下一页</a>
```

```
<a href="<?php echo $href;?>&pageno=<?php echo $pagecount;?>">尾页</a>
<?php
}
?>
[共找到<?php echo $rows;?>个记录]
</div>
</body>
</html>
```

入库单打印页面，3_dayin_rukudan.php 代码：

```
<?php
session_start();
$入库编号=$_SESSION['入库编号'];
  Header("Content-type:application/octet-stream");
  Header("Accept-Ranges:bytes");
  Header("Content-type:application/vnd.ms-excel");
  Header("Content-Disposition:attachment;filename=test.xls");
$link=mysql_connect("127.0.0.1","root","123456")
or die("数据库服务器连接失败！<BR>");
mysql_select_db("kcgl",$link) or die("数据库选择失败！<BR>");
mysql_query("set names 'gbk'");//数据库编码
$sql="select 入库编号,物品编号,物品名称,颜色,大小,进价,货位编号,入库日期,入库数量,负责人 from 入库表 where 入库编号='$入库编号'";
$result=mysql_query($sql,$link);
  echo "入库编号\t物品编号\t物品名称\t颜色\t大小\t进价\t货位编号\t入库日期\t入库数量\t负责人";//输出的excel的列名,用\t换列
    while($rs=mysql_fetch_array($result)){
    echo "\n";//用\n换行
    echo $rs['入库编号']."\t".$rs['物品编号']."\t".$rs['物品名称']."\t".$rs['颜色']."\t".$rs['大小']."\t".$rs['进价']."\t".$rs['货位编号']."\t".$rs['入库日期']."\t".$rs['入库数量']."\t".$rs['负责人'];}
?>
```

代码运行效果见图13.26。

按入库编号查询： 1001　　　查询　取消

导出Excel

入库单查询结果

入库编号	物品编号	物品名称	颜色	大小	进价	货位编号	入库日期	入库数量	负责人
1001	100001	特步女款运动鞋	红色	37	350	111	2014-08-05	12	李
1001	100001	特步女款运动鞋	红色	38	350	111	2014-08-05	12	李
1001	100001	特步女款运动鞋	黄色	38	350	111	2014-08-05	13	李
1001	100002	红蜻蜓男款皮鞋	黑色	40	300	112	2014-08-05	10	李
1001	100002	红蜻蜓男款皮鞋	黑色	41	300	112	2014-08-05	11	李
1001	100002	红蜻蜓男款皮鞋	棕色	41	300	112	2014-08-05	9	李
1001	100002	红蜻蜓男款皮鞋	棕色	42	300	112	2014-08-05	15	李
1001	100003	百丽女款休闲鞋	白色	38	200	113	2014-08-05	10	李
1001	100003	百丽女款休闲鞋	黑色	38	200	113	2014-08-05	11	李

[第1页/共1页] [共找到9个记录]

图13.26 入库单查询

五、考核标准

(1) 版面布局合理清晰,整体效果美观,观赏性强。(10分)

(2) 网页中没有明显的错误(如超链接、图片无法显示、错别字等)。(10分)

(3) 用户登录功能模块。(20分)

(4) 出库管理、入库管理、库存警报。(30分)

(5) 管理员功能模块。(10分)

(6) 新闻公告。(10分)

(7) 创新性、其他功能。(10分)

项目十四　帝　国　CMS

一、实训目的

- 掌握帝国 CMS 的安装配置方法；
- 掌握帝国 CMS 模板制作方法；
- 掌握帝国 CMS 网站制作方法。

二、实训要求

- 使用帝国 CMS 开发商品网站；
- 使用帝国 CMS 进行模板设计。

三、实训设计

对于网站建设和信息发布人员来说，他们最关注的是系统的易用性和功能的完善性，因此，这对网站建设和信息发布工具提出了一个很高的要求。此外，保障网站架构的安全性也是用户关注的焦点。能有效管理网站访问者的登录权限，使内网数据库不受攻击，从而时刻保证网站的安全稳定，免除用户的后顾之忧。

根据以上需求，一套专业的内容管理系统帝国 CMS 应运而生，能有效解决用户网站建设与信息发布中常见的问题和需求。

本实训通过帝国 CMS 的安装配置、模板管理、栏目管理、信息管理，有步骤地培养学生如何快速使用帝国 CMS 开发网站的各个环节，达到快速建站的目标。

四、实训内容

任务一　帝国 CMS 的安装和配置

目前国内常见的 CMS 有很多，我们这里以帝国 CMS 为列来进行网站开发。

首先从 http://www.phome.net/ 下载帝国 CMS 最新版本，然后进行安装。

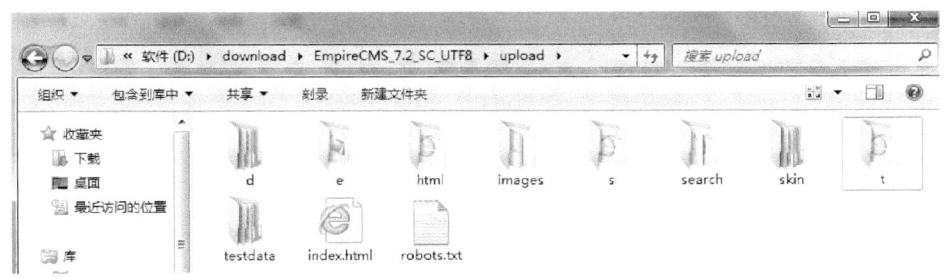

图 14.1　文件复制

(1) 将安装包的 upload 目录中的全部文件和目录结构复制到网站根目录,见图 14.1。

(2) 在浏览器中运行 http://localhost/安装目录/e/install/index.php,进入帝国网站管理系统安装界面,阅读用户使用条款,单击"我同意"按钮,进行运行环境检测,见图 14.2。

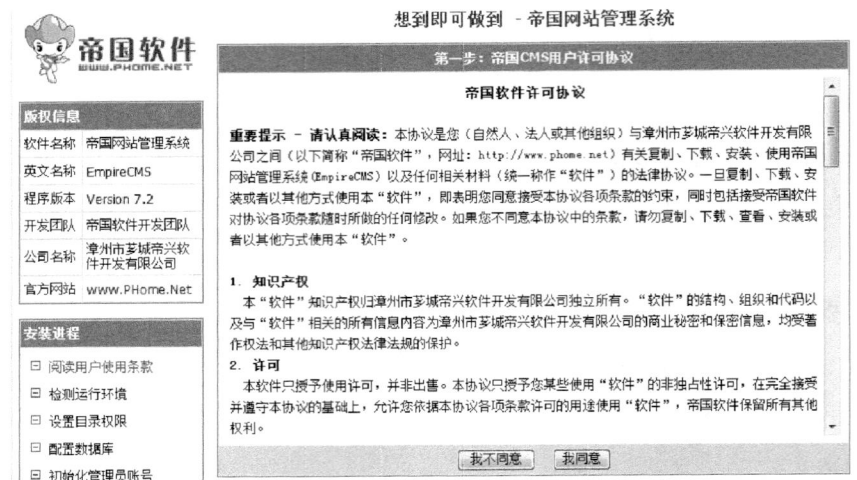

图 14.2　安装界面

(3) 运行环境检查结果通过,单击"下一步"按钮,见图 14.3。

图 14.3　检查环境

(4) 填写 mysql 数据库账号密码,如果需要附带测试数据,则在"内置初始数据"一项打勾,见图 14.4。

(5) 配置好数据库后,单击"下一步"按钮,进行管理员账号设置操作,见图 14.5。

图 14.4 数据库配置

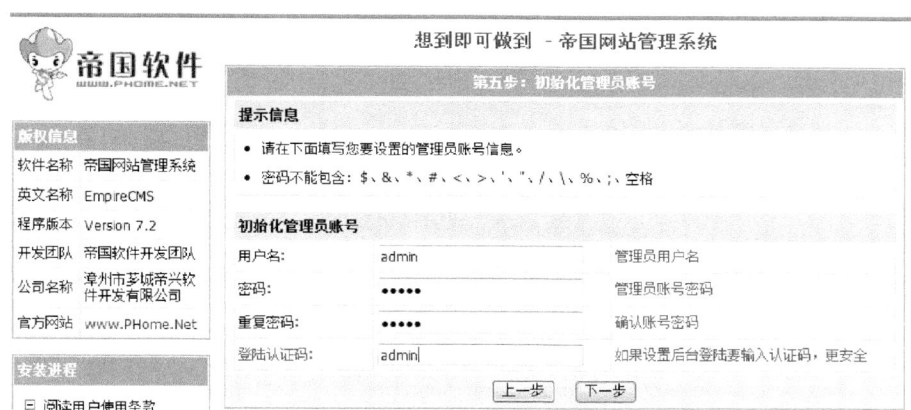

图 14.5 管理员账号设置

(6) 设置好管理员账号后,单击"下一步"按钮,系统安装完毕,见图 14.6。

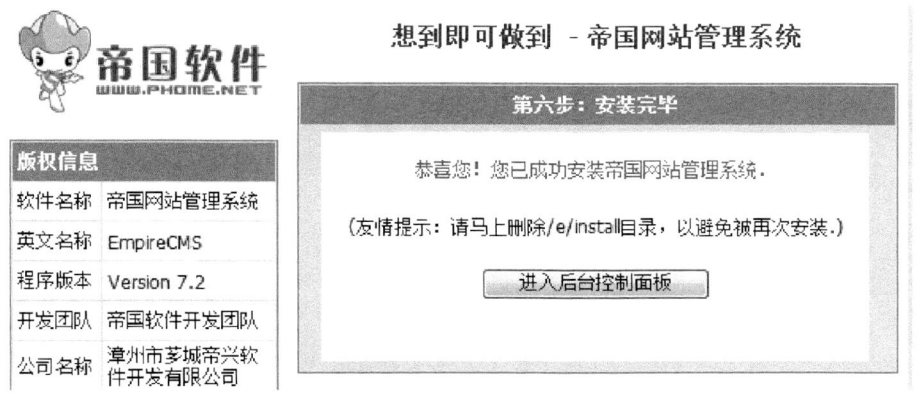

图 14.6 安装完毕

（7）单击"进入后台控制面板"按钮，登录后台，进行系统初始化数据设置，见图 14.7。

图 14.7 登录界面

（8）单击"系统设置"菜单→"数据更新中心"进行初始化默认数据，执行步骤顺序，见图 14.8。

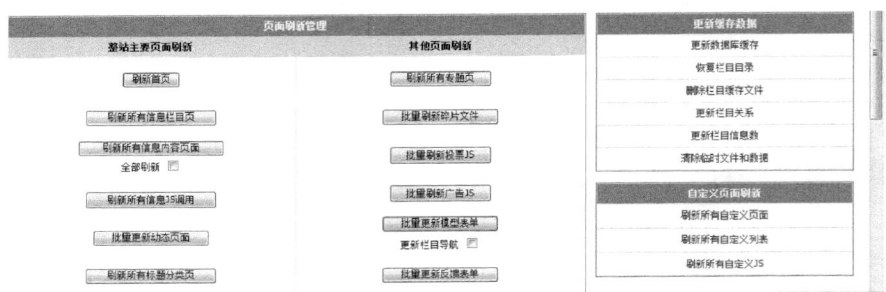

图 14.8 初始化配置

（9）设置站点名称，见图 14.9。

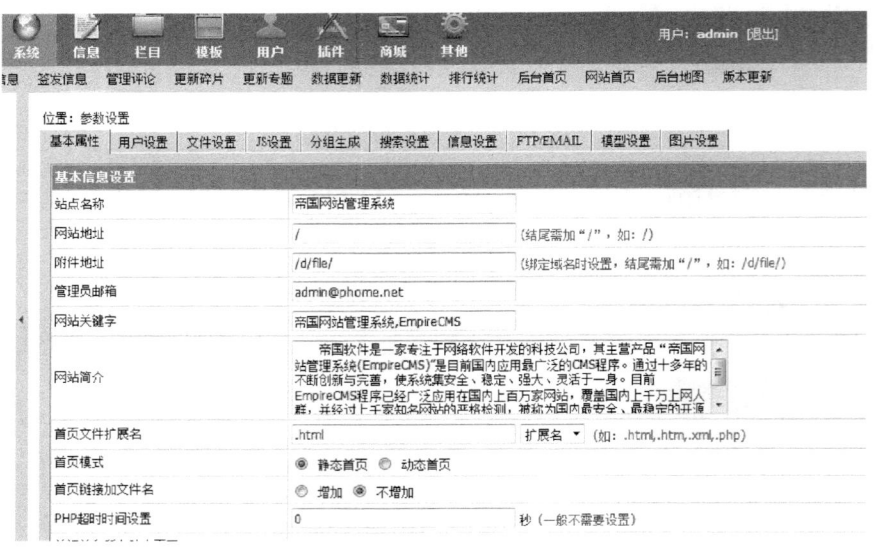

图 14.9 站点设置

到此，帝国 CMS 的安装和配置全部完成。

任务二 模板管理

帝国 CMS 的开发部分主要集中在模板管理中,本模块决定了网站的显示效果。从网站首页到网站栏目页,再到具体子页面,都在此进行开发设置。模板管理页面如图所示,主要包括首页模板、公共模板变量、标签模版、列表模板、内容模板等,见图 14.10。

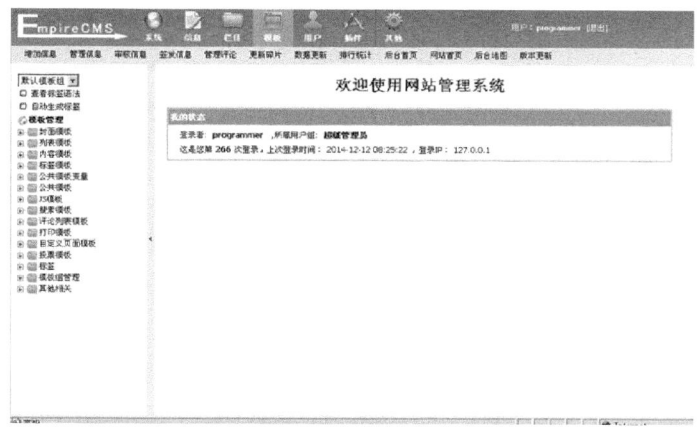

图 14.10 模板管理

1. 首页模板

首页模板决定了首页显示的内容,是所有模板中最重要的部分之一。见图 14.11,首页模板包含在公共模板中,页面代码写在右侧窗体中。

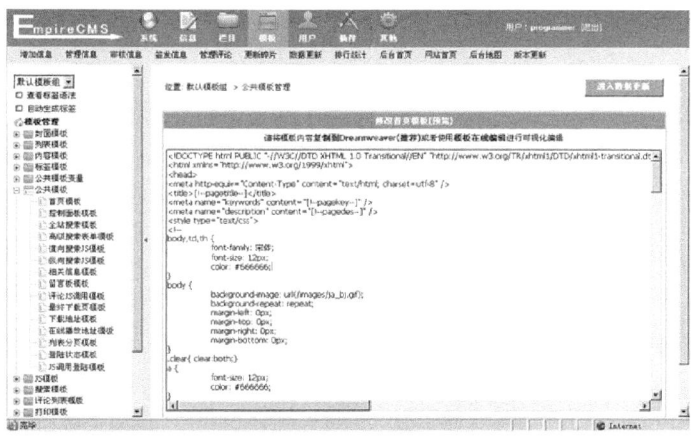

图 14.11 首页模板

首页具体代码如下:

<! DOCTYPE html PUBLIC "-//W3C//DTD XHTML 1.0 Transitional//EN" "http://www.w3.org/TR/xhtml1/DTD/xhtml1-transitional.dtd">
<html xmlns="http://www.w3.org/1999/xhtml">
<head>
<meta http-equiv="Content-Type" content="text/html; charset=utf-8" />
<title>[!--pagetitle--]</title>

```html
<meta name="keywords" content="<!--pagekey-->" />
<meta name="description" content="<!--pagedes-->" />
<style type="text/css">
<!--
body,td,th {
    font-family: 宋体;
    font-size: 12px;
    color: #666666;
}
body {
    background-image: url(/images/ja_bj.gif);
    background-repeat: repeat;
    margin-left: 0px;
    margin-top: 0px;
    margin-right: 0px;
    margin-bottom: 0px;
}
.clear{ clear:both;}
a {
    font-size: 12px;
    color: #666666;
}
a:link {
    text-decoration: none;
}
a:visited {
    text-decoration: none;
    color: #666666;
}
a:hover {
    text-decoration: none;
    color: #F30000;
}
a:active {
    text-decoration: none;
    color: #666666;
}
.top {color: #999999}
.dh {
    color: #F30000;
    font-size: 14px;
}
.dh a {
    font-size: 14px;
    color: #F30000;
}
.dh a:link {
    text-decoration: none;
}
.dh a:visited {
    text-decoration: none;
```

```css
        color:#F30000;
}
.dh a:hover {
    text-decoration:none;
    color:#F30000;
    font-weight:bold;
}
.dh a:active {
    text-decoration:none;
    color:#F30000;
}
.lm {
    color:#333333;
    font-size:14px;
    font-weight:bold;
}
.lb {
    color:#333333;
    font-weight:bold;
}
.dd {color:#cccccc}
-->
</style>
```

```html
<SCRIPT type=text/javascript>kfguin="407606225";ws="http://cs.lsgzn.com/cslsgzn";companyname="兰舍硅藻泥-0731家居网";welcomeword="您好,欢迎光临兰舍硅藻泥<brT>请问,有什么可以帮到您的吗?";type="1";</SCRIPT>
<SCRIPT src="/js/kf.js" type=text/javascript></SCRIPT>
<SCRIPT language=javascript src="/js/check.js"></SCRIPT>
<script type="text/javascript">
function bookmarksite(title,url){ if (document.all)
            window.external.AddFavorite(url, title); else if
            (window.sidebar) window.sidebar.addPanel(title, url, "") }

function setHomepage(url){    // 设为首页
    if (document.all){
        document.body.style.behavior = 'url(#default#homepage)';
        document.body.setHomePage(url);
    }else if (window.sidebar){
        if (window.netscape){
            try {
netscape.security.PrivilegeManager.enablePrivilege("UniversalXPConnect");
            }catch (e) {
                alert("操作被拒绝,请在浏览器地址栏输入 about:config,然后将项 signed.applets.codebase_principal_support 值改为 true");
            }
        }
        var prefs = Components.classes['@mozilla.org/preferences-service;1'].getService(Components.interfaces.nsIPrefBranch);
        prefs.setCharPref('browser.startup.homepage', url);
    }
```

```html
        }
    </script>
    <script type="text/javascript" src="/js/jquery-1.4.2.min.js"></script>
    <script src="/js/jquery.KinSlideshow-1.2.1.min.js" type="text/javascript"></script>
    <script type="text/javascript">
        $(function(){
            $("#KinSlideshow").KinSlideshow();
        })
    </script>
    <script type="text/javascript" src="/js/kxbdMarquee.js"></script>
</head>
<body style="text-align:center">
<table width="100%" border="0" cellspacing="0" cellpadding="0">
  <tr>
    <td height="75" align="center" valign="top">[!--temp.head--]
      <table width="980" border="0" cellspacing="0" cellpadding="0" style="border-left:#DDDDDD solid 1px; border-right:#DDDDDD solid 1px">
        <tr>
          <td height="10"  bgcolor="#FFFFFF"></td>
        </tr>
        <tr>
          <td height="340" align="center" bgcolor="#FFFFFF" valign="top"><table width="100%" height="453" border="0" cellpadding="0" cellspacing="0" style="margin-bottom:20px">
            <tr>
              <td width="200" height="453" align="center" valign="top">[!--temp.left--]</td>
              <td width="778" valign="top" align="left"><table width="768" border="0" cellspacing="0" cellpadding="0">
                <tr>
                  <td width="388" height="119" valign="top" align="left"><table width="380" border="0" cellspacing="0" cellpadding="0">
                    <tr>
                      <td height="32" colspan="2" valign="middle"><table width="100%" border="0" cellspacing="0" cellpadding="0" height="17">
                        <tr>
                          <td width="80" height="17" align="left" valign="middle" background="/images/dd1.jpg" style="background-position:left; background-repeat:no-repeat">  
                            <div  class="position" style="display:inline" rel="p2"><span class="lm" style="line-height:17px">产品知识</span></div></td>
                          <td width="300" height="17" align="left" valign="middle" background="/images/dd.gif"> </td>
                        </tr>
                      </table></td>
                    </tr>
                    <tr>
                      <td width="162" height="99" align="left" valign="middle"><a href="[!--news.url--]chanpinzhishi/" target=_blank title="硅藻泥四大功能    "><img src="/images/1.jpg" width="160" height="120"  border="0" alt="硅藻泥四大功能    "/></a> </td>
                      <td width="218" valign="top"><table width="100%" border="0"
```

```
                    cellspacing="0" cellpadding="0">
                        <tr>
                            [ecmsinfo]'56',5,0,0,0,1[/ecmsinfo]
                        </table></td>
                    </tr>
                </table></td>
                <td width="380" valign="top"><table width="380" border="0" cellspacing="0" cellpadding="0">
                    <tr>
                        <td height="32" colspan="2" valign="middle"><table width="100%" border="0" cellspacing="0" cellpadding="0" height="17">
                            <tr>
                                <td width="80" height="17" align="left" valign="middle" background="/images/dd1.jpg" style="background-position:left;background-repeat:no-repeat">  
                                <div class="position" style="display:inline" rel="p3"><span class="lm" style="line-height:17px">新闻动态</span></div></td>
                                <td width="300" height="17" align="left" valign="middle" background="/images/dd.gif"> </td>
                            </tr>
                        </table></td>
                    </tr>
                    <tr>
                        <td width="162" height="99" align="left" valign="middle"><a href="[!--news.url--]news/" target="_blank" title="兰舍硅藻泥称雄2012年欧亚店庆"><img src="/images/1-12060611244a92.JPG" width="160" height="120" border="0" alt="兰舍硅藻泥称雄2012年欧亚店庆"/></a> </td>
                        <td width="218" valign="top"><table width="100%" border="0" cellspacing="0" cellpadding="0">
                            <tr>
                                [ecmsinfo]'2',5,0,0,0,2[/ecmsinfo]
                            </table></td>
                    </tr>
                </table>
                <strong></strong></td>
            </tr>
            <tr>
                <td height="128" colspan="2"><table width="768" border="0" cellspacing="0" cellpadding="0" style="margin-top:15px">
                    <tr>
                        <td height="22" colspan="2" valign="middle"><table width="100%" border="0" cellspacing="0" cellpadding="0" height="17">
                            <tr>
                                <td width="80" height="17" align="left" valign="middle" background="/images/dd1.jpg" style="background-position:left;background-repeat:no-repeat">  <span class="lm" style="line-height:17px">关于我们</span></td>
                                <td width="688" height="17" align="left" valign="middle" background="/images/dd.gif"> </td>
                            </tr>
                        </table></td>
                    </tr>
```

```html
					<tr>
						<td width="323" height="170" align="left" valign="middle"><img width="300" height="160" alt="" src="/images/201111121105384586.jpg"></td>
						<td width="445" valign="top"><table width="96%" height="166" border="0" cellpadding="0" cellspacing="0">
							<tr>
								<td height="166" style="line-height:20px" valign="center" align="center"><div style="width:420px; height:160px; overflow:hidden;" align="left">
		[!--temp.contactHOME--]...  <a href="[!--news.url--]aboutus.html">更多>></a></div></td>
							</tr>
						</table></td>
					</tr>
				</table></td>
			</tr>
		</table></td>
	</tr>
</table>
   [!--temp.foot--]</td>
 </tr>
</table>
</td>
</tr>
</table>
</body>
</html>
[!--temp.afterbody--]
```

以首页代码为例，我们可以看到帝国CMS代码以静态html代码为主，动态部分使用帝国CMS的模板标签。对于程序员来说，只要会用html设计页面，然后在静态页面中加入帝国CMS的模板标签，就能够制作出动态网站。因此，帝国CMS可以大大降低网站的开发难度，提高开发效率。

模板的调用方法：

[!--temp.name--]

"name"为模板名称，其余为固定部分。如[!-- temp.head --]就是引用名为"temp.head"的模板，那么"temp.head"的内容就会出现标签所在的位置。

在代码中寻找具体的帝国CMS模板标签是比较麻烦的事情，可以点击"模板在线编辑"辅助查看页面中的模板标签。

帝国CMS为了便于初学者了解页面标签，在模板代码下方，提供了快速查阅按钮，包括：显示模板变量说明、查看模板标签语法、查看js调用地址、查看公共模板变量、查看标签模板。图14.12为查看模板变量说明的效果，显示了本页面能够使用的系统模板变量名称。如，[!--pagetitle--]就代表了网站名称。

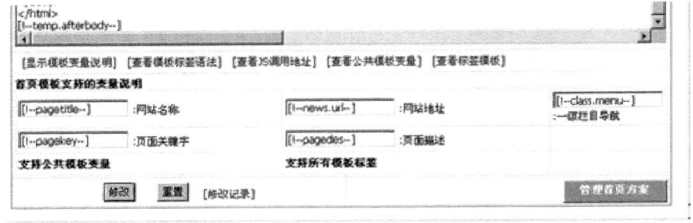

图14.12　首页模板支持的变量

点击"查看模板标签语法",效果见图14.13。可以在弹出页面中,查看帝国CMS所有的标签。

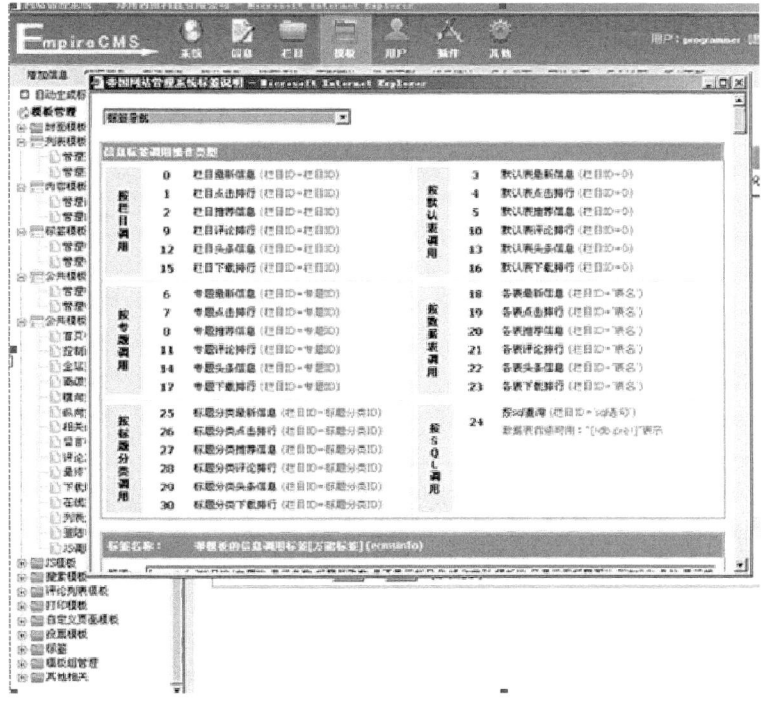

图 14.13　标签语法

点击"查看公共模板变量",效果见图14.14,可以看到所有自己定义的公共模板。

图 14.14　公共模板

点击"查看标签模板",效果见图14.15,可以查看所有定义的标签模板。

2. 公共模板变量

帝国CMS将整个页面拆分为若干部分,每一部分就是一个模板。相当于整个网页由很多名为模

图 14.15 标签模板

板的积木拼成。其中公共模板变量主要是各个页面中都会用到的公共部分代码，比如网站的头部、底部、导航等，这些部分在不同的页面中，基本都是相同的，我们就可以将这部分作为公共模板变量，供其他模板调用。所以，我们也可以这样来理解，公共模板变量是一些小的积木，而上一节讲到的首页模板就是一个整体，而这个整体就是由主体部分（首页模板本身）和各个小积木组成的。

点击页面左侧的公共模板变量——管理模板变量后，进入模板变量管理页面，见图 14.16。

图 14.16 公共模板变量

> [!--temp.head--]头部模板：

```
<table width="980" border="0" cellspacing="0" cellpadding="0" style="border-left:#DDDDDD solid 1px; border-right:#DDDDDD solid 1px">
  <tr>
    <td height="25" colspan="2" align="right" background="/images/ja_top.gif" class="top" valign="middle"><a href="javascript:setHomepage('[!--news.url--]');" _fcksavedurl="javascript:setHomepage('[!--news.url--]');"><span class="top">设为首页</span></a> |
    <a href="javascript:bookmarksite('兰舍硅藻泥','[!--news.url--]')"><span class="top">添
```

```html
加收藏夹</span></a> | <a href="/cslsgzn/contact.html"><span class="top">联系方式</span></a>    </td>
        </tr>
        <tr>
          <td width="410" height="103" background="/images/top1.gif" align="right"><a href="[!--news.url--]"><img src="/images/4fdc564b24b94.jpg" width="387" height="80" border="0" alt="兰舍硅藻泥长沙总代理 环保涂料|兰舍硅藻泥背景墙-长沙有机涂料"/></a></td>
          <td width="570" background="/images/top1.gif"> </td>
        </tr>
      </table>
      <table width="980" border="0" cellspacing="0" cellpadding="0" style="border-left:#DDDDDD solid 1px;border-right:#DDDDDD solid 1px">
        <tr>
          <td height="33" align="center" background="/images/dh1.jpg" class="top"><table width="920" height="33" border="0" cellspacing="0" cellpadding="0">
            <tr>
              <td width="92" align="center" class="dh"><a href="[!--news.url--]">首页</a></td>
              <td width="92" align="center" class="dh"><a href="[!--news.url--]aboutus.html">关于我们</a></td>
              <td width="92" align="center" class="dh"><a href="[!--news.url--]news/" onClick="this.blur()">新闻动态</a></td>
              <td width="92" align="center" class="dh"><a href="[!--news.url--]wenli/" onClick="this.blur()">纹理系列</a></td>
              <td width="92" align="center" class="dh"><a href="[!--news.url--]pinghu/" onClick="this.blur()">平湖系列</a></td>
              <td width="92" align="center" class="dh"><a href="[!--news.url--]chanpinzhishi/" onClick="this.blur()">产品知识</a></td>
              <td width="92" align="center" class="dh"><a href="[!--news.url--]contact.html">联系我们</a></td>
              <td width="92" align="center" class="dh"><a href="[!--news.url--]comment.html">留言板</a></td>
            </tr>
          </table></td>
        </tr>
        <tr>
          <td height="340" align="center" bgcolor="#FFFFFF" valign="middle"><table width="960" height="320" border="0" cellspacing="0" cellpadding="0" style="border:#dddddd solid 1px">
            <tr>
              <td align="left"><div id="KinSlideshow" style="visibility:hidden;"> <a><img src="/images/502b16412793f.jpg" alt="" width="958" height="318" /></a> <a><img src="/images/5023875fc280e.jpg" alt="" width="958" height="318" /></a> </div></td>
            </tr>
          </table></td>
        </tr>
      </table>
      [!--temp.foot--]底部：
      <table width="980" border="0" align="center" cellpadding="0" cellspacing="0" style="border-left:#DDDDDD solid 1px;border-right:#DDDDDD solid 1px;">
        <tr>
          <td align="center" bgcolor="#FFFFFF" valign="middle" style="border-top:#
```

```html
EB000C solid 2px"><table width="100%" border="0" cellspacing="0" cellpadding="0">
    <tr>
        <td width="160" align="center" style="line-height:22px"> </td>
        <td width="660" height="80" align="center" style="line-height:22px">[!--temp.footcontact--]</td>
    <tr> </tr>
    </table></td>
    </tr>
</table>
[!--temp.left--]首页左侧
<table width="180" border="0" cellpadding="0" cellspacing="0" style="border:#dddddd solid 1px">
    <tr>
        <td height="29" background="/images/lm.gif" align="left" valign="middle">  <span class="lm">联系方式</span></td>
    </tr>
    <tr>
        <td height="180" valign="top"><table width="100%" border="0" cellspacing="0" cellpadding="0">
            <tr>
                <td height="180" align="center" style="line-height:24px" valign="top"><div style="width:170px;word-wrap:break-word;padding-top:5px;overflow:hidden;" align="left">[!--temp.leftContact--]</div></td>
            </tr>
        </table></td>
    </tr>
    <tr>
        <td></td>
    </tr>
</table>
```

➢ [!--temp.leftContact--]左侧联系方式：

地址:长沙市万家丽中路与长沙大道交界居然之家高桥店
电话:0731-88725123
传真:0731-88725123
移动电话:13739071508
网址:http://www.cslanshe.com
E-mail:407606225@qq.com
在线客服:<a target="blank" href="http://wpa.qq.com/msgrd?V=1&Uin=407606225"><img src="/images/qqonline.gif" width="21" height="21" border="0" /></a>
<a target="blank" href="http://wpa.qq.com/msgrd?V=1&Uin=407606225"><img src="/images/qqonline.gif" width="21" height="21" border="0" /></a>

➢ [!--temp.contactHOME--]首页关于我们：

兰舍硅藻泥总部位于亚洲最大的硅藻土产地吉林省。公司利用长白山丰富和高品位的硅藻土资源,依托长春高校和科研单位在硅藻土研究方面取得最新成果,兰舍硅藻泥系列产品已渐成为行业翘楚。公司最近推出的新产品"硅藻泥+贝壳粉"套餐彻底解决了困扰和限制硅藻泥行业发展的施工和价格瓶颈,为硅藻泥这一"旧时王谢堂前燕"能够"飞入寻常百姓家"铺平了道路。

➢ [!--temp.lefts--]其他页面左侧:

```html
<td width="200" height="453" align="center" valign="top"><table width="180"  border="0" cellpadding="0" cellspacing="0" style="border:#dddddd solid 1px">
    <tr>
      <td height="29" background="/images/lm.gif" align="left" valign="middle">  <span class="lm">新闻动态</span></td>
    </tr>
    <tr>
      <td  height="180"  valign="top"><table width="100%" border="0" cellspacing="0" cellpadding="0">
          <tr>
            <td  height="30" align="left" background="/images/xx.jpg" style="background-position:bottom; background-repeat:no-repeat">  <img src="/images/jia.gif" width="7" height="7" />
                <div style="overflow:hidden; width:100px; height:28px; display:inline">  <a href="[!--news.url--]news/qiyexinwen/"><span class="lb">企业新闻</span></a></div></td>
          </tr>
          <tr>
            <td  height="30" align="left" background="/images/xx.jpg" style="background-position:bottom; background-repeat:no-repeat">  <img src="/images/jia.gif" width="7" height="7" />
                <div style="overflow:hidden; width:100px; height:28px; display:inline">  <a href="[!--news.url--]news/xingyexinwen/"><span class="lb">行业新闻</span></a></div></td>
          </tr>
      </table></td>
    </tr>
    <tr>
      <td></td>
    </tr>
</table>
<table width="180"  border="0" cellpadding="0" cellspacing="0" style="border:#dddddd solid 1px; margin-top:10px">
    <tr>
      <td height="29" background="http://cs.lsgzn.com/application/views/template/vip/jiuai/jiuai/lm.gif" align="left" valign="middle">  <span class="lm">联系方式</span></td>
    </tr>
    <tr>
      <td  height="180"  valign="top" align="center"><table width="96%" border="0" cellspacing="0" cellpadding="0">
          <tr>
            <td  height="180" align="center" style="line-height:24px" valign="top"><div style="width:170px;  padding-top:5px; word-break:break-all" align="left">[!--temp.leftContact--]</div></td>
          </tr>
      </table></td>
    </tr>
    <tr>
      <td height="2"></td>
    </tr>
```

```
</table></td>
```

> [!--temp.footcontact--]底部联系方式：

```
<p align="center"><span style="font-family:宋体;">长沙家居网</span><span style="font-family:宋体;"> mailto:0731jiaju@longwing.cn<br />
       copyright 2007 longwing.cn all rights reserved.<br />
治理甲醛  净化空气  消除异味 调节湿度 防潮防霉  防火隔热  <br />
       </span></p>
       <p align="center"><span style="font-family:宋体;">兰舍硅藻泥长沙总代理|长沙环保涂料|长沙背景墙|长沙有机涂料|长沙硅藻泥<br />
       <br />
       长沙市万家丽中路长沙大道口6楼1015号</span></p>
       <p align="center"><span style="font-family:宋体;">东方新城H4栋108门面(万家丽观光梯侧50米)</span></p>
       <script type="text/javascript">
var _bdhmProtocol = (("https:" == document.location.protocol) ? " https://" : " http://");
document.write(unescape("%3Cscript src='" + _bdhmProtocol + "hm.baidu.com/h.js%3F184f90d3ab4a0f85ea220a451af0cbf3' type='text/javascript'%3E%3C/script%3E"));
</script>
       技术支持:<a href="http://www.0731jiaju.com.cn" target="_blank">0731家居网</a>
```

3. 标签模板

标签是帝国 CMS 自定义的,具有动态输出效果的特殊标记。

标签模板主要控制帝国 CMS 各个标签的显示效果。换句话说,标签控制了输出什么,而标签模板控制了输出效果。点击页面左侧标签模板——管理标签模版,效果见图 14.17,显示了所有标签模板。

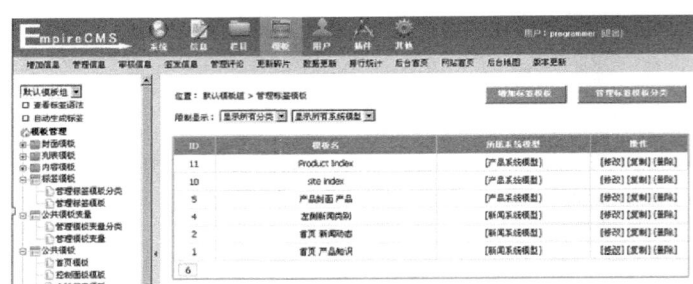

图 14.17 标签模板

> 首页产品知识：

从图 14.18 可以看到,在标签模板中,代码部分分为"页面模板内容"和"列表内容模板"。其中"页面模板内容"调用了"列表内容模板"。比如新闻,在首页中一次要输出多条,我们不可能在模板中也写多条输出语句,而是用循环的办法进行多次输出。而"列表内容模板"就是一次输出的效果,而"页面模板内容"则调用"列表内容模板",并最终根据具体的参数在网页中进行多次输出。

页面模板内容代码：

[!--empirenews.listtemp--]<!--list.var1-->[!--empirenews.listtemp--]

列表内容模板代码：

图 14.18 产品知识

```
<td height="26"><div style="overflow:hidden; width:100%; height:26px; line-height:26px">
<span class="dd">·</span><a href="[!--titleurl--]">[!--title--]</a></div>
</td>
</tr>
```

首页 新闻动态：

```
[!--empirenews.listtemp--]<!--list.var1-->[!--empirenews.listtemp--]
<td height="26"><div style="overflow:hidden; width:100%; height:26px; line-height:26px">
<span class="dd">·</span><a href="[!--titleurl--]">[!--title--]</a></div>
</td>
</tr>
```

➢ 左侧新闻类别模板：

页面模板内容代码：

```
[!--empirenews.listtemp--]<!--list.var1-->[!--empirenews.listtemp--]
```

列表内容模板代码：

```
<tr>
<td height="30" align="left" background="/images/xx.jpg" style="background-position:bottom; background-repeat:no-repeat">  <img src="/images/jia.gif" width="7" height="7" />
<div style="overflow:hidden; width:100px; height:28px; display:inline">  <a href="[!--titleurl--]"><span class="lb">[!--this.classname--]</span></a></div>
</td>
</tr>
```

➢ 产品封面模板：

页面模板内容代码：

```
[!--empirenews.listtemp--]
<tr>
<!--list.var1-->
<!--list.var2-->
<!--list.var3-->
<!--list.var4-->
```

```
        </tr>
        [!--empirenews.listtemp--]
```

列表内容模板代码:

```
<td align="center" valign="middle" style="line-height:28px; padding-left:10px; padding-right:10px"><table width="150" height="180" border="0" align="center" cellpadding="0" cellspacing="0">
    <tr>
        <td height="150" align="center" valign="middle" style="border:#dddddd solid 1px"><a href="[!--titleurl--]" target="_blank"><img src="[!--titlepic--]" width="150" height="150" border="0" onload="javascript:DrawImage(this,150,150)" id="img" name="img" /></a></td>
    </tr>
    <tr>
        <td height="30" align="center" valign="bottom" class="lm_12"><a href="[!--titleurl--]" target="_blank">[!--title--]</a><br/>
        </td>
    </tr>
</table>
<p></p></td>
```

4. 列表模板

列表模板主要负责栏目的页面内容。比如新闻栏目,打开栏目后,看到的页面就是由列表模板控制的。也可以将这个页面称为栏目首页,不同的栏目如果首页相似,则可以使用同一个列表模板。

点击页面左侧列表模板——管理列表模板,如图 14.19 所示,可以看到所有的列表模板。

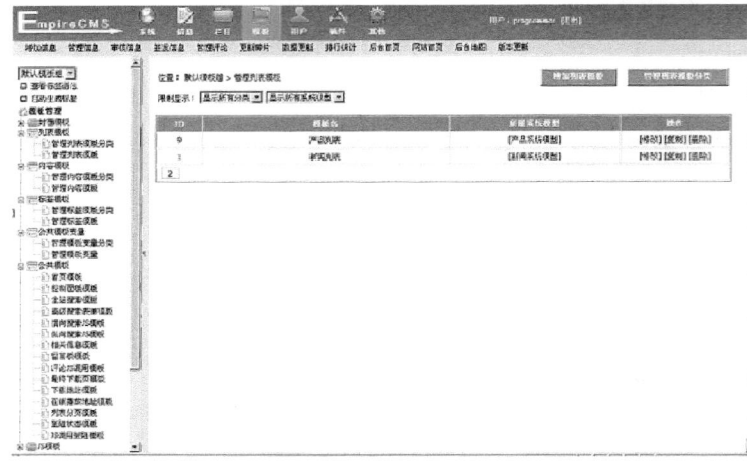

图 14.19 列表模板

➤ 新闻列表:

页面模板内容代码:

```
<!DOCTYPE html PUBLIC "-//W3C//DTD XHTML 1.0 Transitional//EN" "http://www.w3.org/TR/xhtml1/DTD/xhtml1-transitional.dtd">
<html xmlns="http://www.w3.org/1999/xhtml">
<head>
<meta http-equiv="Content-Type" content="text/html; charset=utf-8" />
```

```html
<title>[!--pagetitle--]</title>
<meta name="keywords" content="[!--pagekeywords--]" />
<meta name="description" content="[!--pagedescription--]" />
<style type="text/css">
<!--
body,td,th {
    font-family: 宋体;
    font-size: 12px;
    color: #666666;
}
body {
    background-image: url(/images/ja_bj.gif);
    background-repeat: repeat;
    margin-left: 0px;
    margin-top: 0px;
    margin-right: 0px;
    margin-bottom: 0px;
}
a {
    font-size: 12px;
    color: #666666;
}
a:link {
    text-decoration: none;
}
a:visited {
    text-decoration: none;
    color: #666666;
}
a:hover {
    text-decoration: none;
    color: #F30000;
}
a:active {
    text-decoration: none;
    color: #666666;
}
.top {color: #999999}
.dh {
    color: #F30000;
    font-size: 14px;
}
.dh a {
    font-size: 14px;
    color: #F30000;
}
.dh a:link {
    text-decoration: none;
}
.dh a:visited {
    text-decoration: none;
```

```css
        color:#F30000;
}
.dh a:hover {
        text-decoration:none;
        color:#F30000;
        font-weight:bold;
}
.dh a:active {
        text-decoration:none;
        color:#F30000;
}
.lm {
        color:#333333;
        font-size:14px;
        font-weight:bold;
}
.lb {
        color:#333333;
        font-weight:bold;
}
.dd {color:#cccccc}
-->
</style>
```

```html
<SCRIPT type=text/javascript>kfguin="407606225";ws="http://cs.lsgzn.com/cslsgzn";companyname="兰舍硅藻泥-0731家居网";welcomeword="您好,欢迎光临兰舍硅藻泥<brT>请问,有什么可以帮到您的吗?";type="1";</SCRIPT>
<SCRIPT src="/js/kf.js" type=text/javascript></SCRIPT>
<SCRIPT language=javascript src="/js/check.js"></SCRIPT>
<script type="text/javascript">
function bookmarksite(title,url){ if (document.all)
        window.external.AddFavorite(url, title); else if
        (window.sidebar) window.sidebar.addPanel(title, url, "") }

function setHomepage(url){    // 设为首页
    if (document.all){
        document.body.style.behavior = 'url(#default#homepage)';
        document.body.setHomePage(url);
    }else if (window.sidebar){
        if (window.netscape){
            try {
                netscape.security.PrivilegeManager.enablePrivilege("UniversalXPConnect");
            }catch (e) {
                alert("操作被拒绝,请在浏览器地址栏输入 about:config,然后将项 signed.applets.codebase_principal_support 值改为 true");
            }
        }
        var prefs = Components.classes['@mozilla.org/preferences-service;1'].getService(Components.interfaces.nsIPrefBranch);
        prefs.setCharPref('browser.startup.homepage', url);
    }
```

```
            }
        </script>
        <script type="text/javascript" src="/js/jquery-1.4.2.min.js"></script>
        <script src="/js/jquery.KinSlideshow-1.2.1.min.js" type="text/javascript"></script>
        <script type="text/javascript">
            $(function(){
                $("#KinSlideshow").KinSlideshow();
            })
        </script>
    </head>
    <body style="text-align:center">
        <table width="100%" border="0" cellspacing="0" cellpadding="0">
            <tr>
                <td height="75" align="center" valign="top">[!--temp.head--]
                    <table width="980" border="0" cellspacing="0" cellpadding="0" style="border-left:#DDDDDD solid 1px; border-right:#DDDDDD solid 1px">
                        <tr>
                            <td height="10"  bgcolor="#FFFFFF"></td>
                        </tr>
                        <tr>
                            <td height="340" align="center" bgcolor="#FFFFFF" valign="top"><table width="100%" height="453" border="0" cellpadding="0" cellspacing="0" style="margin-bottom:20px">
                                <tr>[!--temp.lefts--]
                                    <td width="778" valign="top" align="left"><table width="768" border="0" cellspacing="0" cellpadding="0">
                                        <tr>
                                            <td height="475" colspan="2" valign="top"><table width="768" border="0" cellspacing="0" cellpadding="0" >
                                                <tr>
                                                    <td height="22" valign="middle"><table width="100%" border="0" cellspacing="0" cellpadding="0" height="17">
                                                        <tr>
                                                            <td width="80" height="17" align="left" valign="middle" background="/images/dd1.jpg" style="background-position:left; background-repeat:no-repeat" >  <span class="lm" style="line-height:17px">[!--class.name--]</span></td>
                                                            <td width="688" height="17" align="left" valign="middle" background="/images/dd.gif"> </td>
                                                        </tr>
                                                    </table></td>
                                                </tr>
                                                <tr>
                                                    <td height="453" align="center" valign="top"><table width="96%" height="424" border="0" cellpadding="0" cellspacing="0">
                                                        <tr>
                                                            <td height="424" style=" padding-top:10px; line-height:22px" valign="top" align="left"><table width="96%" height="424" border="0" cellpadding="0" cellspacing="0">
                                                                <tr>
                                                                    <td height="424" style=" padding-top:10px" valign
```

```html
="top" align="center"><table width="699" border="0" cellpadding="0" cellspacing="0">
<tr>
<td width="699" height="28" valign="top" class="about" style="padding-bottom:10px;">[!--empirenews.listtemp--]
<!--list.var1-->
[!--empirenews.listtemp--]
<table width="98%" border="0" align="center" cellpadding="0" cellspacing="0">
<tr>
<td height="30" align="right" valign="center" style="letter-spacing:2px;">[!--show.page--]</td>
</tr>
</table></td>
</tr>
</table></td>
</tr>
</table></td>
</tr>
</table></td>
</tr>
</table></td>
</tr>
</table>
[!--temp.foot--]
</td>
</tr>
</table>
</body>
</html>
[!--temp.afterbody--]
```

列表内容模板代码：

```html
<table width="98%" height="30" border="0" cellpadding="0" cellspacing="0" class="bg1_new" onMouseOver="this.className='bg2_new'" onMouseOut="this.className='bg1_new'" style="border-bottom:#999999 dotted 1px" align="center">
<tr>
<td width="3%" align="center" valign="middle"><span style="color:#FF9900">·</span></td>
<td width="71%" align="left" valign="middle"><div style="width:100%; height:24px; overflow:hidden; text-align:left; line-height:24px"><a href="[!--titleurl--]">[!--title--]</a></div></td>
<td width="26%" align="right" valign="middle"><span style="font-family:Arial, Helvetica, sans-serif; color:#958b6c; font-size:11px;">[!--newstime--]    </span></td>
```

```
        </tr>
    </table>
```

> 新闻列表：

页面模板内容代码：

```
<!DOCTYPE html PUBLIC "-//W3C//DTD XHTML 1.0 Transitional//EN" "http://www.w3.org/TR/xhtml1/DTD/xhtml1-transitional.dtd">
<html xmlns="http://www.w3.org/1999/xhtml">
<head>
<meta http-equiv="Content-Type" content="text/html; charset=utf-8" />
<title>[!--pagetitle--]</title>
<meta name="keywords" content="[!--pagekey--]" />
<meta name="description" content="[!--pagedes--]" />
<style type="text/css">
<!--
body, td, th {
    font-family: 宋体;
    font-size: 12px;
    color: #666666;
}
body {
    background-image: url(/images/ja_bj.gif);
    background-repeat: repeat;
    margin-left: 0px;
    margin-top: 0px;
    margin-right: 0px;
    margin-bottom: 0px;
}
a {
    font-size: 12px;
    color: #666666;
}
a:link {
    text-decoration: none;
}
a:visited {
    text-decoration: none;
    color: #666666;
}
a:hover {
    text-decoration: none;
    color: #F30000;
}
a:active {
    text-decoration: none;
    color: #666666;
}
.top {color: #999999}
.dh {
    color: #F30000;
```

```css
        font-size: 14px;
}
.dh a {
        font-size: 14px;
        color: #F30000;
}
.dh a:link {
        text-decoration: none;
}
.dh a:visited {
        text-decoration: none;
        color: #F30000;
}
.dh a:hover {
        text-decoration: none;
        color: #F30000;
        font-weight:bold;
}
.dh a:active {
        text-decoration: none;
        color: #F30000;
}
.lm {
        color: #333333;
        font-size: 14px;
        font-weight: bold;
}
.lb {
        color: #333333;
        font-weight: bold;
}
.dd {color: #cccccc}
.hot_pro {width:735px;}
.hot_pro ul{overflow:hidden;zoom:1;}
.hot_pro   p{line-height:14px;}
.hot_pro li{float:left;width:140px;height:140px; display:block; margin:10px 0 0 15px;}
.hot_pro li .pic{position: relative;display: block;overflow: hidden; width:140x; height:105px;}
-->
</style>
<SCRIPT type=text/javascript> kfguin = "407606225"; ws = "http://cs.lsgzn.com/cslsgzn"; companyname="兰舍硅藻泥-0731家居网"; welcomeword="您好,欢迎光临兰舍硅藻泥<brT>请问,有什么可以帮到您的吗?"; type="1";</SCRIPT>
<SCRIPT src="/js/kf.js" type=text/javascript></SCRIPT>
<SCRIPT language=javascript src="/js/check.js"></SCRIPT>
<script type="text/javascript">
function bookmarksite(title, url){ if (document.all)
            window.external.AddFavorite(url, title); else if
            (window.sidebar) window.sidebar.addPanel(title, url, "") }

function setHomepage(url){    // 设为首页
    if (document.all){
```

```
                document.body.style.behavior = 'url(#default#homepage)';
                document.body.setHomePage(url);
            }else if (window.sidebar){
                if (window.netscape){
                    try {
                        netscape.security.PrivilegeManager.enablePrivilege("UniversalXPConnect");
                    }catch (e) {
                        alert("操作被拒绝,请在浏览器地址栏输入 about:config,然后将项 signed.applets.codebase_principal_support 值改为 true");
                    }
                }
                var prefs = Components.classes['@mozilla.org/preferences-service;1'].getService(Components.interfaces.nsIPrefBranch);
                prefs.setCharPref('browser.startup.homepage', url);
            }
        }
    </script>
    <script type="text/javascript" src="/js/jquery-1.4.2.min.js"></script>
    <script src="/js/jquery.KinSlideshow-1.2.1.min.js" type="text/javascript"></script>
    <script type="text/javascript">
    $(function(){
        $("#KinSlideshow").KinSlideshow();
    })
    </script>
</head>
<body style="text-align:center">
<table width="100%" border="0" cellspacing="0" cellpadding="0">
  <tr>
    <td height="75" align="center" valign="top">[!--temp.head--]
      <table width="980" border="0" cellspacing="0" cellpadding="0" style="border-left:#DDDDDD solid 1px; border-right:#DDDDDD solid 1px">
        <tr>
          <td height="10"   bgcolor="#FFFFFF"></td>
        </tr>
        <tr>
          <td height="340" align="center" bgcolor="#FFFFFF" valign="top"><table width="100%" height="453" border="0" cellpadding="0" cellspacing="0" style="margin-bottom:20px">
            <tr>[!--temp.ProLeft--]
              <td width="778" valign="top" align="left"><table width="768" border="0" cellspacing="0" cellpadding="0">
                <tr>
                  <td height="475" colspan="2" valign="top"><table width="768" border="0" cellspacing="0" cellpadding="0" >
                    <tr>
                      <td height="22" valign="middle"><table width="100%" border="0" cellspacing="0" cellpadding="0" height="17">
                        <tr>
                          <td width="100" height="17" align="left" valign="middle" background="/images/dd1.jpg" style="background-position:left; background-repeat:no-repeat"
```

```html
>  <span class="lm" style="line-height:17px">[!--class.name--]</span>
</td>
                                  <td width="688" height="17" align="left" valign="middle" background="/images/dd.gif"> </td>
                                </tr>
                              </table></td>
                            </tr>
                            <tr>
                              <td height="453" align="center" valign="top"><table width="96%" height="424" border="0" cellpadding="0" cellspacing="0">
                                <tr>
                                  <td height="424" style=" padding-top:10px; line-height:22px" valign="top" align="left"><table width="96%" height="424" border="0" cellpadding="0" cellspacing="0">
                                    <tr>
                                      <td height="424" style=" padding-top:10px" valign="top" align="center"><table width="699"  border="0" cellpadding="0" cellspacing="0">
                                        <tr>
                                          <td width="699" height="28" valign="top" class="about" style="padding-bottom:10px;"><div class="hot_pro">
        <ul>
          [!--empirenews.listtemp--]
          <!--list.var1-->
          [!--empirenews.listtemp--]
        </ul>
      </div>
                                            <table width="98%" border="0" align="center" cellpadding="0" cellspacing="0">
                                              <tr>
                                                <td height="30" align="right" valign="center"  style="letter-spacing:2px;">[!--show.page--]</td>
                                              </tr>
                                            </table></td>
                                        </tr>
                                      </table></td>
                                    </tr>
                                  </table></td>
                                </tr>
                              </table></td>
                            </tr>
                          </table></td>
                        </tr>
                      </table></td>
                    </tr>
                  </table></td>
                </tr>
              </table>
              [!--temp.foot--]
            </td>
```

```
</tr>
</table>
</body>
</html>
[!--temp.afterbody--]
```

列表内容模板代码：

```
<li><a href="[!--titleurl--]" target="_blank"><img src="[!--titlepic--]" width="150" height="150" border="0" onload="javascript:DrawImage(this,150,150)" id="img" name="img" /></a>
<p><a href="[!--titleurl--]" target="_blank"><center>[!--title--]</center></a></p>
</li>
```

在列表管理页面，我们还可以看到管理系统模型按钮，点击后可以看到系统的系统模型，见图14.20。

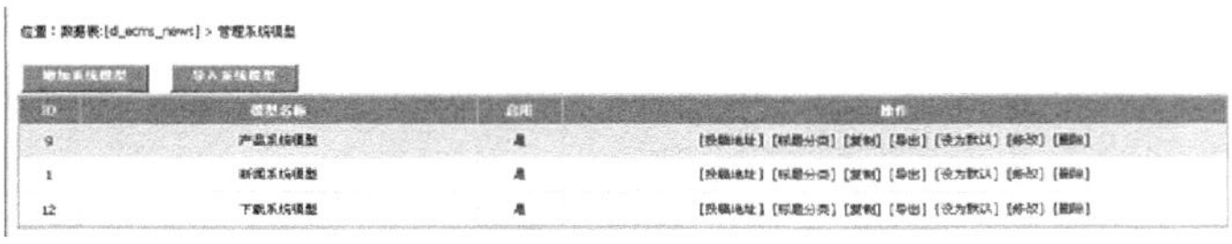

图 14.20 系统模型

系统模型基于数据库中的表格生成，点击修改按钮可以看到，见图14.21。

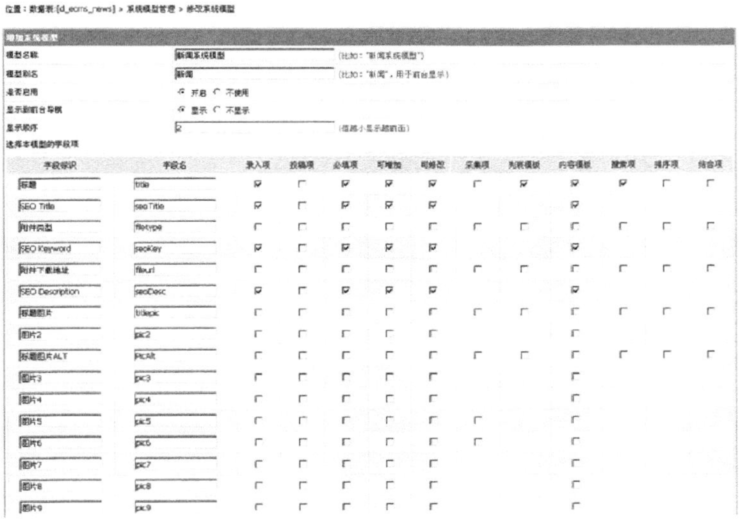

图 14.21 模型选项

在创建列表模板的时候，必须选择系统模型，实际上也是选择这个列表模板使用了数据库中的哪个表格，决定了最终数据将会在哪个表格中进行存取。

5. 内容模板

内容模板决定了最终的子页面显示效果，如打开一条具体的新闻时看到的页面。同列表模板一样，如果两个子页面显示效果相似，可以使用同一个内同模板。如，新闻可以分类，但是不同类型的新闻除了新闻内同，其他都是一样的，就叫做页面显示效果相似。

点击页面左侧内容模板——管理内容模板,如图 14.22 所示,就可以看到所有的内容模板。

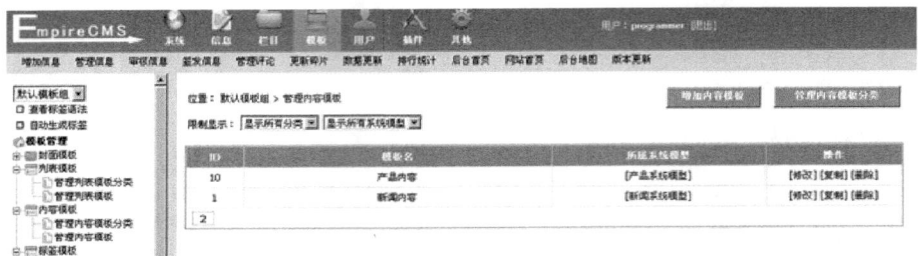

图 14.22 内容模板

➤ 新闻内容:

```
<!DOCTYPE html PUBLIC "-//W3C//DTD XHTML 1.0 Transitional//EN" "http://www.w3.org/TR/xhtml1/DTD/xhtml1-transitional.dtd">
<html xmlns="http://www.w3.org/1999/xhtml">
<head>
<meta http-equiv="Content-Type" content="text/html; charset=utf-8" />
<title>[!--seoTitle--]</title>
<meta name="keywords" content="[!--seoKey--]" />
<meta name="description" content="[!--seoDesc--]" />
<style type="text/css">
<!--
body, td, th {
    font-family: 宋体;
    font-size: 12px;
    color: #666666;
}
body {
    background-image: url(/images/ja_bj.gif);
    background-repeat: repeat;
    margin-left: 0px;
    margin-top: 0px;
    margin-right: 0px;
    margin-bottom: 0px;
}
a {
    font-size: 12px;
    color: #666666;
}
a:link {
    text-decoration: none;
}
a:visited {
    text-decoration: none;
    color: #666666;
}
a:hover {
    text-decoration: none;
    color: #F30000;
```

```css
}
a:active {
    text-decoration: none;
    color: #666666;
}
.top {color: #999999}
.dh {
    color: #F30000;
    font-size: 14px;
}
.dh a {
    font-size: 14px;
    color: #F30000;
}
.dh a:link {
    text-decoration: none;
}
.dh a:visited {
    text-decoration: none;
    color: #F30000;
}
.dh a:hover {
    text-decoration: none;
    color: #F30000;
    font-weight: bold;
}
.dh a:active {
    text-decoration: none;
    color: #F30000;
}
.lm {
    color: #333333;
    font-size: 14px;
    font-weight: bold;
}
.lb {
    color: #333333;
    font-weight: bold;
}
.dd {color: #cccccc}
-->
</style>
<SCRIPT type=text/javascript> kfguin = "407606225"; ws = "http://cs.lsgzn.com/cslsgzn"; companyname = "兰舍硅藻泥-0731家居网"; welcomeword = "您好,欢迎光临兰舍硅藻泥<brT>请问,有什么可以帮到您的吗?"; type = "1";</SCRIPT>
<SCRIPT src = "/js/kf.js" type = text/javascript></SCRIPT>
<SCRIPT language = javascript src = "/js/check.js"></SCRIPT>
<script type = "text/javascript">
function bookmarksite(title, url){ if (document.all)
            window.external.AddFavorite(url, title); else if
            (window.sidebar) window.sidebar.addPanel(title, url, "") }
```

```
function setHomepage(url){    // 设为首页
    if (document.all){
        document.body.style.behavior = 'url(#default#homepage)';
        document.body.setHomePage(url);
    }else if (window.sidebar){
        if (window.netscape){
            try {
                netscape.security.PrivilegeManager.enablePrivilege("UniversalXPConnect");
            }catch (e) {
                alert("操作被拒绝,请在浏览器地址栏输入 about:config,然后将项 signed.applets.codebase_principal_support 值改为 true");
            }
        }
        var prefs = Components.classes['@mozilla.org/preferences-service;1'].getService(Components.interfaces.nsIPrefBranch);
        prefs.setCharPref('browser.startup.homepage', url);
    }
}
</script>
<script type="text/javascript" src="/js/jquery-1.4.2.min.js"></script>
<script src="/js/jquery.KinSlideshow-1.2.1.min.js" type="text/javascript"></script>
<script type="text/javascript">
$(function(){
    $("#KinSlideshow").KinSlideshow();
})
</script>
</head>
<body style="text-align:center">
<table width="100%" border="0" cellspacing="0" cellpadding="0">
  <tr>
    <td height="75" align="center" valign="top">[!--temp.head--]
      <table width="980" border="0" cellspacing="0" cellpadding="0" style="border-left:#DDDDDD solid 1px; border-right:#DDDDDD solid 1px">
        <tr>
          <td height="10" bgcolor="#FFFFFF"></td>
        </tr>
        <tr>
          <td height="340" align="center" bgcolor="#FFFFFF" valign="top"><table width="100%" height="453" border="0" cellpadding="0" cellspacing="0" style="margin-bottom:20px">
            <tr>[!--temp.lefts--]
              <td width="778" valign="top" align="left"><table width="768" border="0" cellspacing="0" cellpadding="0">
                <tr>
                  <td height="475" colspan="2" valign="top"><table width="768" border="0" cellspacing="0" cellpadding="0" >
                    <tr>
                      <td height="22" valign="middle"><table width="100%" border="0" cellspacing="0" cellpadding="0" height="17">
```

```html
            <tr>
                <td width="80" height="17" align="left" valign="middle" background="/images/dd1.jpg" style="background-position:left; background-repeat:no-repeat">  <span class="lm" style="line-height:17px">详细信息</span></td>
                <td width="688" height="17" align="left" valign="middle" background="/images/dd.gif"> </td>
            </tr>
        </table></td>
    </tr>
    <tr>
        <td height="453" align="center" valign="top"><table width="96%" height="424" border="0" cellpadding="0" cellspacing="0">
            <tr>
                <td height="424" style=" padding-top:10px" valign="top" align="center"><table width="94%" border="0" cellpadding="0" cellspacing="0">
                    <tr>
                        <td height="35" align="center" style="border-bottom:dashed #987D44 1px"><table width="100%" height="25" border="0" cellpadding="0" cellspacing="0">
                            <tr>
                                <td height="22"  align="center"><span style="font-size:14px; font-weight:bold">[!--title--]</span> [!--newstime--]
                                <p> </p></td>
                            </tr>
                        </table></td>
                    </tr>
                    <tr>
                        <td  valign="top" align="center"><table width="100%" border="0" align="center" cellpadding="10" cellspacing="0">
                            <tr>
                                <td align="left" valign="middle" style="line-height:20px;"><div class="content">[!--newstext--] </div>
                                <!-- /content -->
                                </td>
                            </tr>
                        </table></td>
                    </tr>
                </table></td>
            </tr>
        </table></td>
    </tr>
</table>
[!--temp.foot--]
```

```
        </td>
    </tr>
</table>
</body>
</html>
[!--temp.afterbody--]
```

效果见图 14.23。

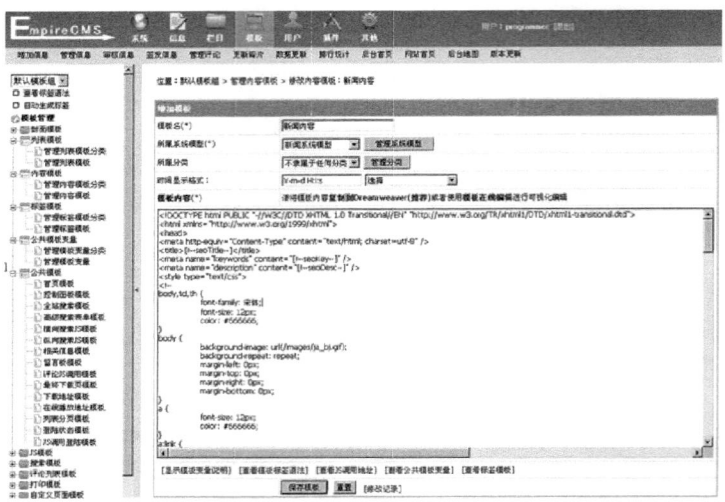

图 14.23　新闻内容模板

➢ 产品代码：

```
<!DOCTYPE html PUBLIC "-//W3C//DTD XHTML 1.0 Transitional//EN" "http://www.w3.org/TR/xhtml1/DTD/xhtml1-transitional.dtd">
<html xmlns="http://www.w3.org/1999/xhtml">
<head>
<meta http-equiv="Content-Type" content="text/html; charset=utf-8" />
<title>[!--seoTitle--]</title>
<meta name="keywords" content="[!--seoKey--]" />
<meta name="description" content="[!--seoDesc--]" />
<style type="text/css">
<!--
body, td, th {
    font-family: 宋体;
    font-size: 12px;
    color: #666666;
}
body {
    background-image: url(/images/ja_bj.gif);
    background-repeat: repeat;
    margin-left: 0px;
    margin-top: 0px;
    margin-right: 0px;
    margin-bottom: 0px;
}
```

```css
a {
    font-size: 12px;
    color: #666666;
}
a:link {
    text-decoration: none;
}
a:visited {
    text-decoration: none;
    color: #666666;
}
a:hover {
    text-decoration: none;
    color: #F30000;
}
a:active {
    text-decoration: none;
    color: #666666;
}
.top {color: #999999}
.dh {
    color: #F30000;
    font-size: 14px;
}
.dh a {
    font-size: 14px;
    color: #F30000;
}
.dh a:link {
    text-decoration: none;
}
.dh a:visited {
    text-decoration: none;
    color: #F30000;
}
.dh a:hover {
    text-decoration: none;
    color: #F30000;
    font-weight: bold;
}
.dh a:active {
    text-decoration: none;
    color: #F30000;
}
.lm {
    color: #333333;
    font-size: 14px;
    font-weight: bold;
}
.lb {
    color: #333333;
```

```
            font-weight: bold;
        }
        .dd {color: #cccccc}
        -->
        </style>
        <SCRIPT type=text/javascript> kfguin = "407606225"; ws = "http://cs.lsgzn.com/cslsgzn"; companyname="兰舍硅藻泥-0731家居网"; welcomeword="您好,欢迎光临兰舍硅藻泥<brT>请问,有什么可以帮到您的吗?"; type="1";</SCRIPT>
        <SCRIPT src="/js/kf.js" type=text/javascript></SCRIPT>
        <SCRIPT language=javascript src="/js/check.js"></SCRIPT>
        <script type="text/javascript">
        function bookmarksite(title, url){ if (document.all)
                window.external.AddFavorite(url, title); else if
                (window.sidebar) window.sidebar.addPanel(title, url, "") }

        function setHomepage(url){    // 设为首页
            if (document.all){
                document.body.style.behavior = 'url(#default#homepage)';
                document.body.setHomePage(url);
            }else if (window.sidebar){
                if (window.netscape){
                    try {
        netscape.security.PrivilegeManager.enablePrivilege("UniversalXPConnect");
                    }catch (e) {
                        alert("操作被拒绝,请在浏览器地址栏输入 about:config,然后将项 signed.applets.codebase_principal_support 值改为 true");
                    }
                }
                var prefs = Components.classes['@mozilla.org/preferences-service;1'].getService(Components.interfaces.nsIPrefBranch);
                prefs.setCharPref('browser.startup.homepage', url);
            }
        }
        </script>
        <script type="text/javascript" src="/js/jquery-1.4.2.min.js"></script>
        <script src="/js/jquery.KinSlideshow-1.2.1.min.js" type="text/javascript"></script>
        <script type="text/javascript">
        $(function(){
            $("#KinSlideshow").KinSlideshow();
        })
        </script>
        <!--以下为焦点图所需代码-->
        <style type="text/css">
        .slide-trigger {
            position:absolute;
            width:400px;
            text-align:right;
            padding-right:5px;
            height:25px;
            z-index:10;
```

```css
}
.slide-trigger a {
    display:inline-block;
    margin-right:3px;
    width:16px;
    height:16px;
    line-height:16px;
    text-align:center;
    color:#d94b01;
    background-color:#fff5e1;
    border:1px solid #f47500;
    outline:none;
    overflow:hidden;
}
.slide-trigger a:hover { text-decoration:none; }
.slide-trigger a.current {
    width:18px;
    height:18px;
    line-height:18px;
    font-weight:bold;
    color:#FFF;
    background:url(/images/t-bg.png) repeat-x;
}
.slide-panel {
    /* 下面四项必须设置 */
    position:relative;
    width:400px;
    height:400px;
    overflow:hidden;
    border:0px;
}
.slide-panel div { position:absolute; bottom:0px; left:0px }
```

```html
</style>
<script type="text/javascript" src="/js/common.js"></script>
<script type="text/javascript">
$(function(){
    if($(".slide-panel > div > table").length > 0){
        window.api = $(".slide-trigger").switchable(".slide-panel > div > table", {
            effect: "scroll",
            vertical: true
        }).carousel().autoplay({ api: true });
    }
});
</script>

</head>
<body style="text-align:center">
<table width="100%" border="0" cellspacing="0" cellpadding="0">
  <tr>
    <td height="75" align="center" valign="top">[!--temp.head--]
```

```html
<table width="980" border="0" cellspacing="0" cellpadding="0" style="border-left:#DDDDDD solid 1px; border-right:#DDDDDD solid 1px">
    <tr>
        <td height="10" bgcolor="#FFFFFF"></td>
    </tr>
    <tr>
        <td height="340" align="center" bgcolor="#FFFFFF" valign="top"><table width="100%" height="453" border="0" cellpadding="0" cellspacing="0" style="margin-bottom:20px">
            <tr>[!--temp.lefts--]
                <td width="778" valign="top" align="left"><table width="768" border="0" cellspacing="0" cellpadding="0">
                    <tr>
                        <td height="475" colspan="2" valign="top"><table width="768" border="0" cellspacing="0" cellpadding="0" >
                            <tr>
                                <td height="22" valign="middle"><table width="100%" border="0" cellspacing="0" cellpadding="0" height="17">
                                    <tr>
                                        <td width="80" height="17" align="left" valign="middle" background="/images/dd1.jpg" style="background-position:left; background-repeat:no-repeat">  <span class="lm" style="line-height:17px">详细信息</span></td>
                                        <td width="688" height="17" align="left" valign="middle" background="/images/dd.gif"> </td>
                                    </tr>
                                </table></td>
                            </tr>
                            <tr>
                                <td height="453" align="center" valign="top"><table width="96%" height="424" border="0" cellpadding="0" cellspacing="0">
                                    <tr>
                                        <td height="424" style=" padding-top:10px" valign="top" align="center"><table width="94%" border="0" cellpadding="0" cellspacing="0">
                                            <tr>
                                                <td height="35" align="center" style="border-bottom:dashed #987D44 1px"><table width="100%" height="25" border="0" cellpadding="0" cellspacing="0">
                                                    <tr>
                                                        <td height="22"  align="center"><span style="font-size:14px; font-weight:bold">[!--title--]</span> [!--newstime--]
                                                        </td>
                                                    </tr>
                                                </table></td>
                                            </tr>
                                            <tr>
                                                <td  valign="top" align="center"><br />
                                                    <table width="386" height="305" border="0" cellpadding="0" cellspacing="0">
                                                        <tr align="middle">
                                                            <td align="center" valign="middle"><div class="slide-panel">
                                                                    <div style="position:absolute">
                                                                        <table width="400" height="400" border
```

```
= "0" cellpadding = "0" cellspacing = "0">
                                        <tr>
                                          <td align = "center" valign = "middle"
><a href = "[!--titlepic--]" target = "_blank"><img src = "[!--titlepic--]" border = "0"
onload = "javascript:DrawImage(this,400,400)" style = "border:none" /></a></td>
                                        </tr>
                                      </table>
                                    </div>
                                    <div class = "slide-trigger">
                                      <!--自动创建 triggers-->
                                    </div>
                                  </div></td>
                                </tr>
                              </table>
                              <table width = "100%" border = "0" cellspacing = "0"
cellpadding = "10">
                                <tr>
                                  <td align = "left" style = "line-height:20px;">
<h3 class = "headline-2 bk-sidecatalog-title">
                                      [!--newstext--]</td>
                                  </tr>
                                </table></td>
                              </tr>
                            </table></td>
                          </tr>
                        </table></td>
                      </tr>
                    </table></td>
                  </tr>
                </table></td>
              </tr>
            </table></td>
          </tr>
        </table>
        [!--temp.foot--]
      </td>
    </tr>
  </table>
  </body>
</html>
[!--temp.afterbody--]
```

效果见图 14.24。

6. 自定义模板

自定义模板主要控制自定义页面的显示效果,自定义页面的内容可以在后台直接修改,通常固定不变,并且不包含子页面。比如网站中的"关于我们",通常情况下"关于我们"的内容是不变的,因此没有必要做成全动态显示的页面。但是,也要考虑到可能存在的变化,就作为自定义页面存在,见图 14.25。

点击修改后看到的效果见图 14.26。

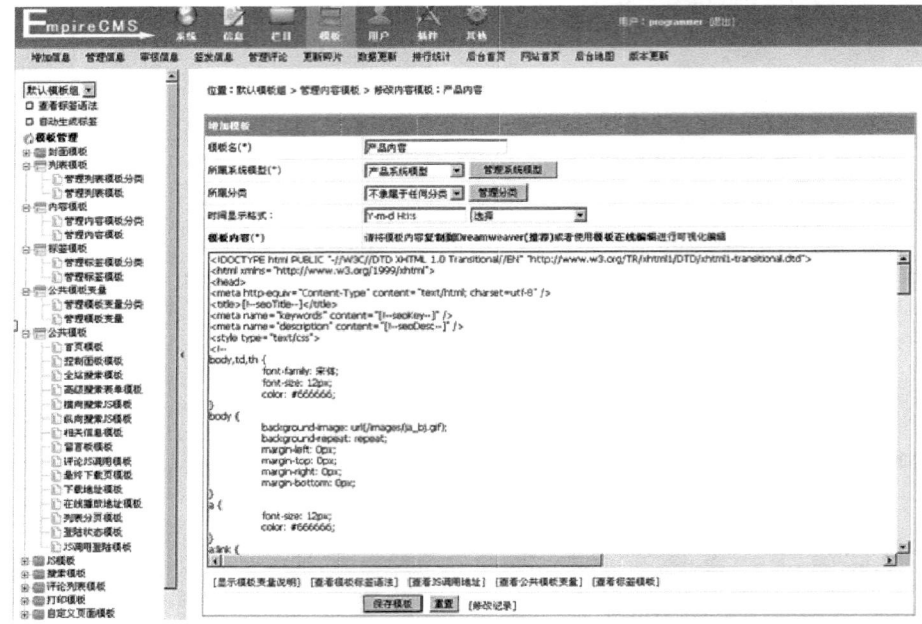

图 14.24　产品内容模板

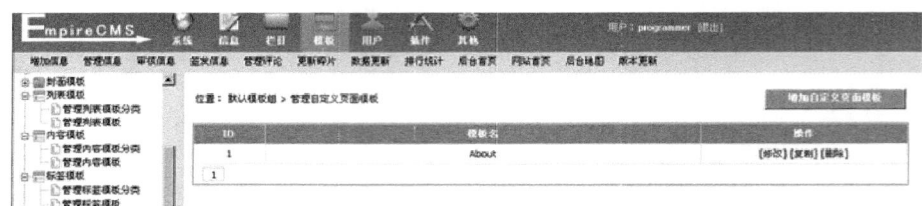

图 14.25　自定义模板

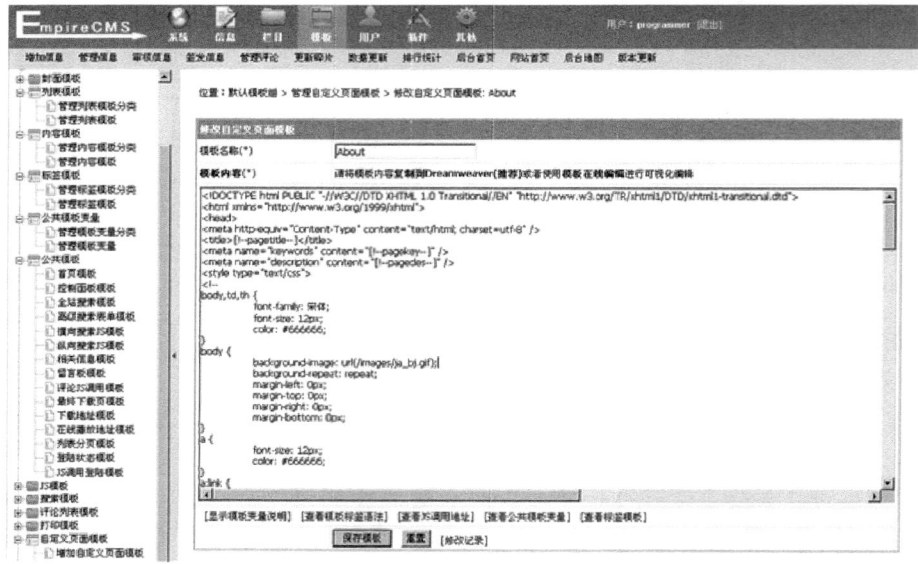

图 14.26　自定义模板内容

其中代码如下：

```html
<!DOCTYPE html PUBLIC "-//W3C//DTD XHTML 1.0 Transitional//EN" "http://www.w3.org/TR/xhtml1/DTD/xhtml1-transitional.dtd">
<html xmlns="http://www.w3.org/1999/xhtml">
<head>
<meta http-equiv="Content-Type" content="text/html; charset=utf-8" />
<title>[!--pagetitle--]</title>
<meta name="keywords" content="[!--pagekey--]" />
<meta name="description" content="[!--pagedes--]" />
<style type="text/css">
<!--
body, td, th {
    font-family: 宋体;
    font-size: 12px;
    color: #666666;
}
body {
    background-image: url(/images/ja_bj.gif);
    background-repeat: repeat;
    margin-left: 0px;
    margin-top: 0px;
    margin-right: 0px;
    margin-bottom: 0px;
}
a {
    font-size: 12px;
    color: #666666;
}
a:link {
    text-decoration: none;
}
a:visited {
    text-decoration: none;
    color: #666666;
}
a:hover {
    text-decoration: none;
    color: #F30000;
}
a:active {
    text-decoration: none;
    color: #666666;
}
.top {color: #999999}
.dh {
    color: #F30000;
    font-size: 14px;
}
.dh a {
    font-size: 14px;
```

```css
        color:#F30000;
    }
    .dh a:link {
        text-decoration:none;
    }
    .dh a:visited {
        text-decoration:none;
        color:#F30000;
    }
    .dh a:hover {
        text-decoration:none;
        color:#F30000;
        font-weight:bold;
    }
    .dh a:active {
        text-decoration:none;
        color:#F30000;
    }
    .lm {
        color:#333333;
        font-size:14px;
        font-weight:bold;
    }
    .lb {
        color:#333333;
        font-weight:bold;
    }
    .dd {color:#cccccc}
    -->
</style>
<SCRIPT type=text/javascript>kfguin="407606225";ws="http://cs.lsgzn.com/cslsgzn";companyname="兰舍硅藻泥-0731家居网";welcomeword="您好,欢迎光临兰舍硅藻泥<brT>请问,有什么可以帮到您的吗?";type="1";</SCRIPT>
<SCRIPT src="/js/kf.js" type=text/javascript></SCRIPT>
<SCRIPT language=javascript src="/js/check.js"></SCRIPT>
<script type="text/javascript">
function bookmarksite(title,url){ if (document.all)
            window.external.AddFavorite(url, title); else if
            (window.sidebar) window.sidebar.addPanel(title, url, "") }

function setHomepage(url){     // 设为首页
        if (document.all){
            document.body.style.behavior = 'url(#default#homepage)';
            document.body.setHomePage(url);
        }else if (window.sidebar){
            if (window.netscape){
                try {
                    netscape.security.PrivilegeManager.enablePrivilege("UniversalXPConnect");
                }catch (e) {
                    alert("操作被拒绝,请在浏览器地址栏输入 about:config,然后将项 signed.applets.
```

```
codebase_principal_support 值改为 true");
                }
            }
            var prefs = Components.classes['@mozilla.org/preferences-service;1'].getService
(Components.interfaces.nsIPrefBranch);
            prefs.setCharPref('browser.startup.homepage',url);
        }
    }
</script>
<script type="text/javascript" src="/js/jquery-1.4.2.min.js"></script>
<script src="/js/jquery.KinSlideshow-1.2.1.min.js" type="text/javascript"></script>
<script type="text/javascript">
$(function(){
    $("#KinSlideshow").KinSlideshow();
})
</script>
</head>
<body style="text-align:center">
<table width="100%" border="0" cellspacing="0" cellpadding="0">
  <tr>
    <td height="75" align="center" valign="top">[!--temp.head--]
      <table width="980" border="0" cellspacing="0" cellpadding="0" style="border-left:#DDDDDD solid 1px; border-right:#DDDDDD solid 1px">
        <tr>
          <td height="10" bgcolor="#FFFFFF"></td>
        </tr>
        <tr>
          <td height="340" align="center" bgcolor="#FFFFFF" valign="top"><table width="100%" height="453" border="0" cellpadding="0" cellspacing="0" style="margin-bottom:20px">
            <tr>[!--temp.lefts--]
              <td width="778" valign="top" align="left"><table width="768" border="0" cellspacing="0" cellpadding="0">
                <tr>
                  <td height="475" colspan="2" valign="top"><table width="768" border="0" cellspacing="0" cellpadding="0" >
                    <tr>
                      <td height="22" valign="middle"><table width="100%" border="0" cellspacing="0" cellpadding="0" height="17">
                        <tr>
                          <td width="80" height="17" align="left" valign="middle" background="/images/dd1.jpg" style="background-position:left; background-repeat:no-repeat">  <span class="lm" style="line-height:17px">[!--pagename--]</span></td>
                          <td width="688" height="17" align="left" valign="middle" background="/images/dd.gif"> </td>
                        </tr>
                      </table></td>
                    </tr>
                    <tr>
                      <td height="453" align="center" valign="top"><table width="
```

```
96%" height="424" border="0" cellpadding="0" cellspacing="0">
                              <tr>
                                <td height="424" style=" padding-top:10px; line-height:
22px" valign="top" align="left">  [!--pagetext--]</td>
                              </tr>
                            </table></td>
                          </tr>
                        </table></td>
                      </tr>
                    </table></td>
                  </tr>
                </table></td>
              </tr>
            </table></td>
          </tr>
        </table>
        [!--temp.foot--]
      </td>
    </tr>
  </table>
</body>
</html>
[!--temp.afterbody--]
```

任务三 栏目管理

模板创建完成之后,就可以开始创建栏目,包括根栏目和子栏目,一般不出现三级目录。栏目管理页面见图 14.27。

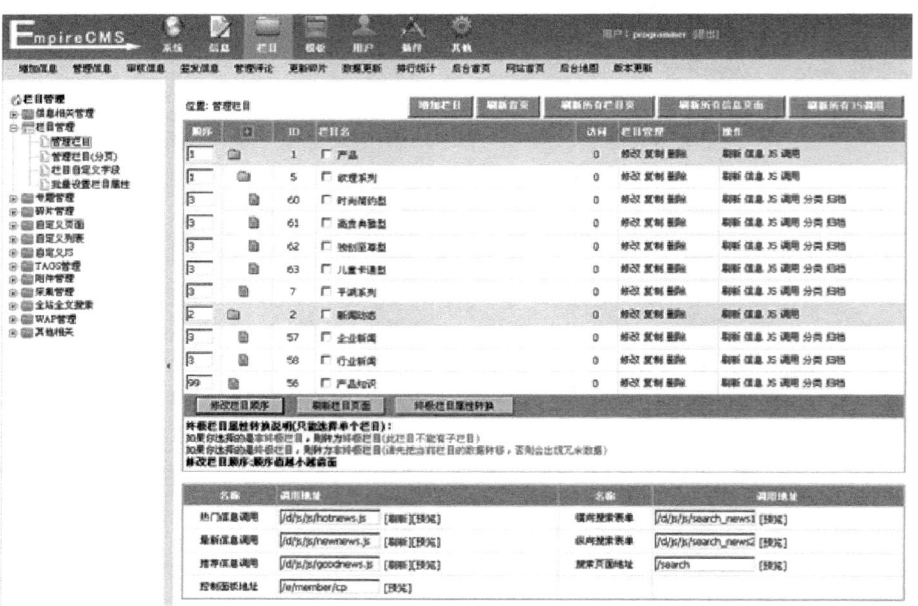

图 14.27 栏目管理

点击"增加栏目"可以创建新的栏目,点击"修改"可以修改栏目。

如图 14.28 所示,创建"纹理系列"栏目。

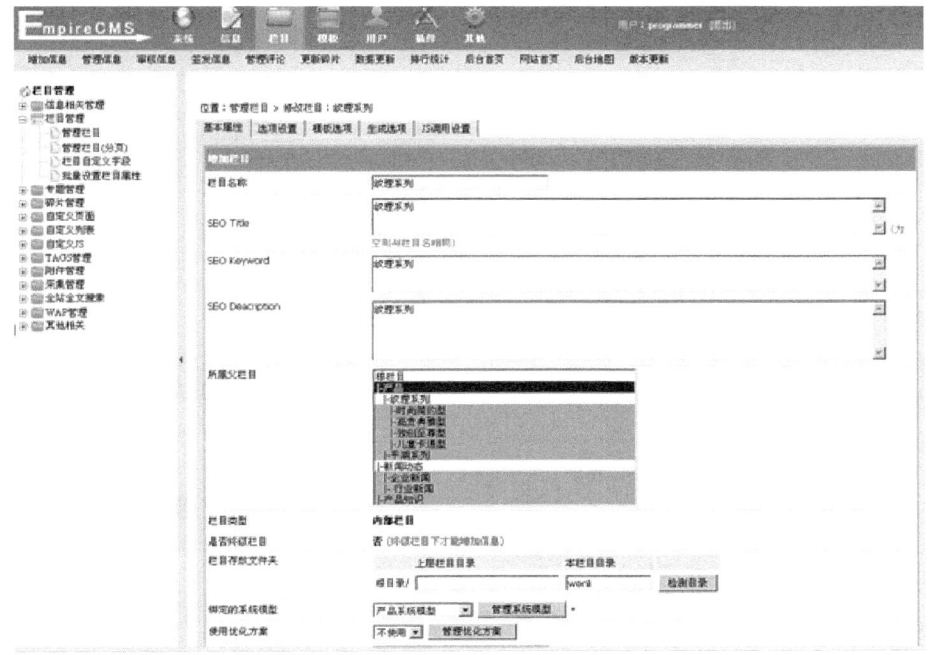

图 14.28　增加栏目

因为是根栏目,也就是顶级栏目,所以"是否终极栏目"一项选择"否",根目录为空,本栏目目录填写栏目名称,不同栏目之间不得重名。绑定的系统模型一项选择要使用的系统模型。

在模板选项中,选择要使用的列表模板,见图 14.29。

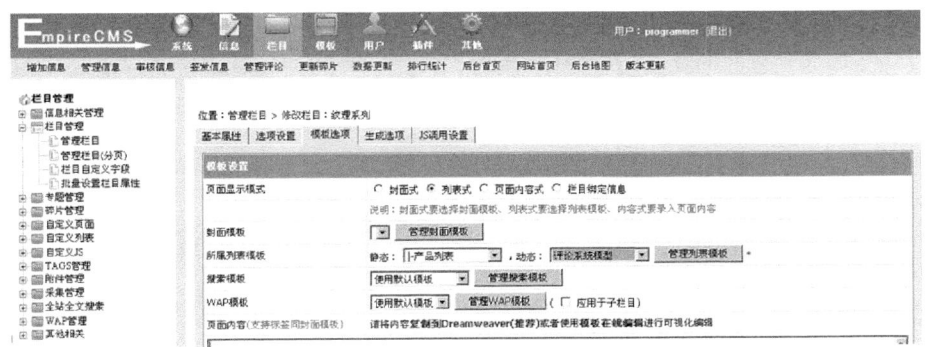

图 14.29　选择列表模板

在生成选项中,选择生成静态页面,见图 14.30。

其他内容默认,完成栏目生成。

添加根栏目之后,可以在根栏目下添加子栏目,如图 14.31 所示时尚简约型。

因为是终极栏目,也就是栏目下不再包含子栏目,所以"是否终极栏目"一项选择"是"。因为是子栏目,所以上层栏目目录填写父栏目名称。

模板选项中选择列表模板和内容模板,见图 14.32。

项目十四　帝国 CMS / 237

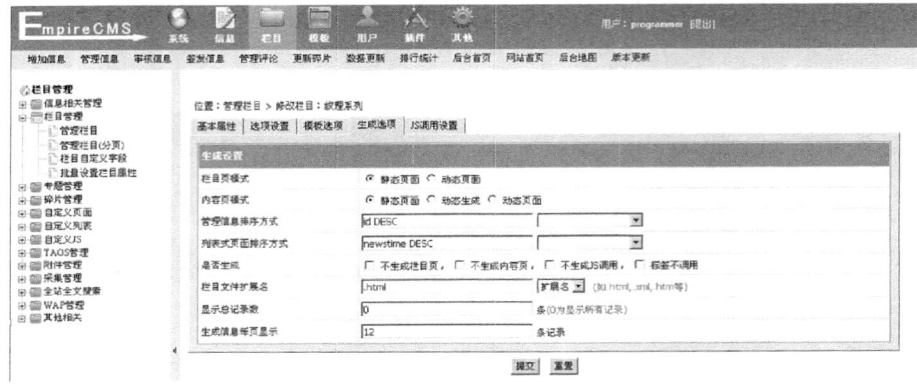

图 14.30　选择生成静态页面

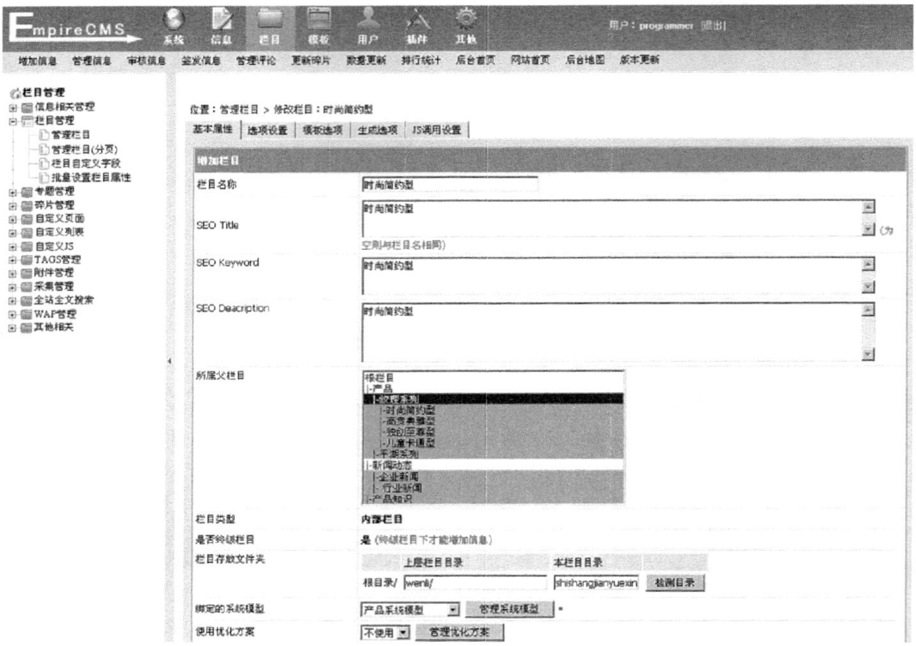

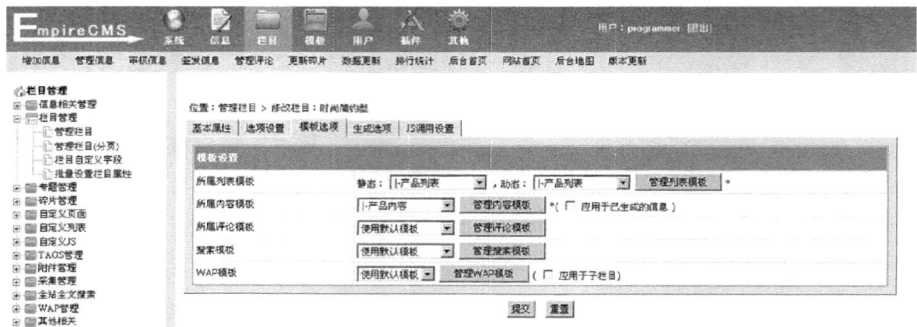

图 14.31　添加子栏目

图 14.32　基本属性

生成选项中选择静态页面,其他默认,见图 14.33。

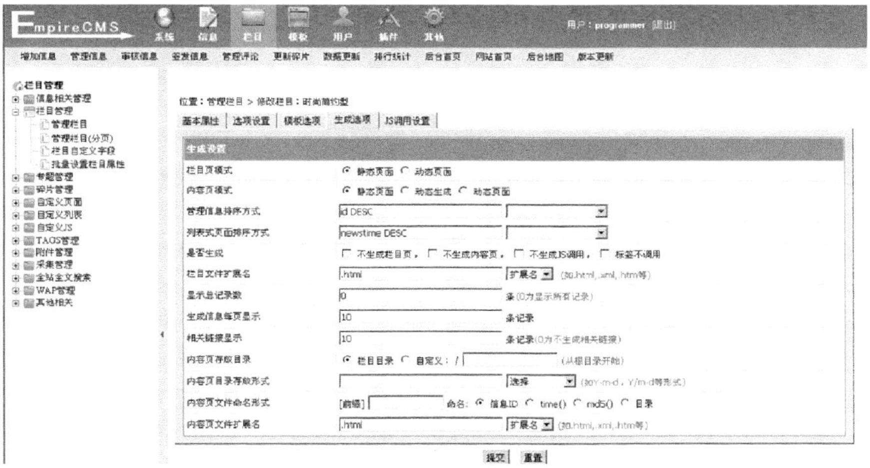

图 14.33 生成选项

因为子栏目之间比较相似,可以点击复制按钮,省略设置过程,见图 14.34。

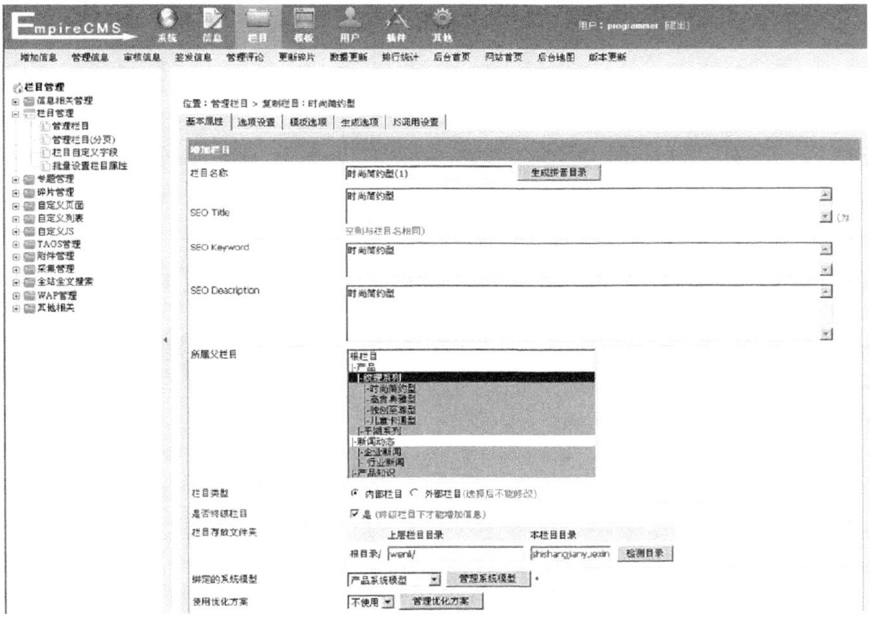

图 14.34 复制栏目

自定义页面依据自定义模板生成,具体页面见图 14.35。

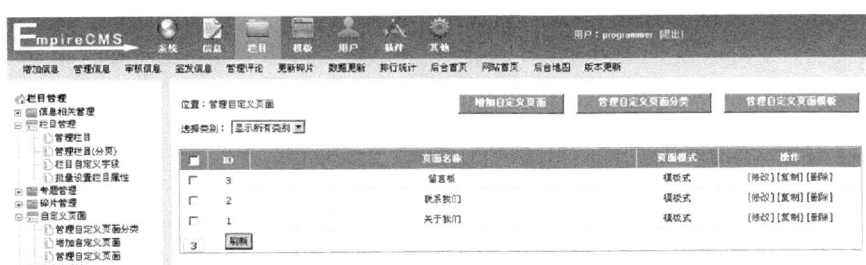

图 14.35 自定义页面

点击修改,见图 14.36。

图 14.36　修改自定义页面

任务四　信 息 管 理

栏目生成后,就可以添加具体的信息,如新闻信息、产品信息等,见图 14.37。

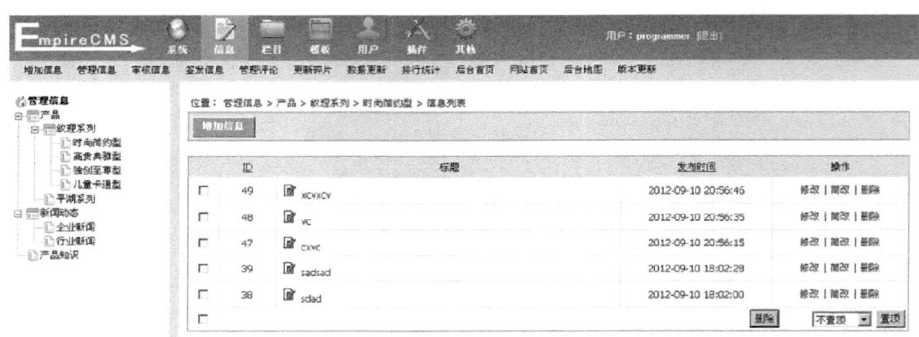

图 14.37　信息管理

新闻编辑页面主要用于编辑新闻:

新闻添加成功后,在系统刷新会刷新,然后就可以在前台页面看到我们编辑的页面了,见图 14.39。

五、考核标准

(1) 版面布局合理清晰,整体效果美观,观赏性强。(10 分)

(2) 网页中没有明显的错误(如超链接、图片无法显示、错别字等)。(10 分)

(3) 帝国 CMS 安装配置。(10 分)

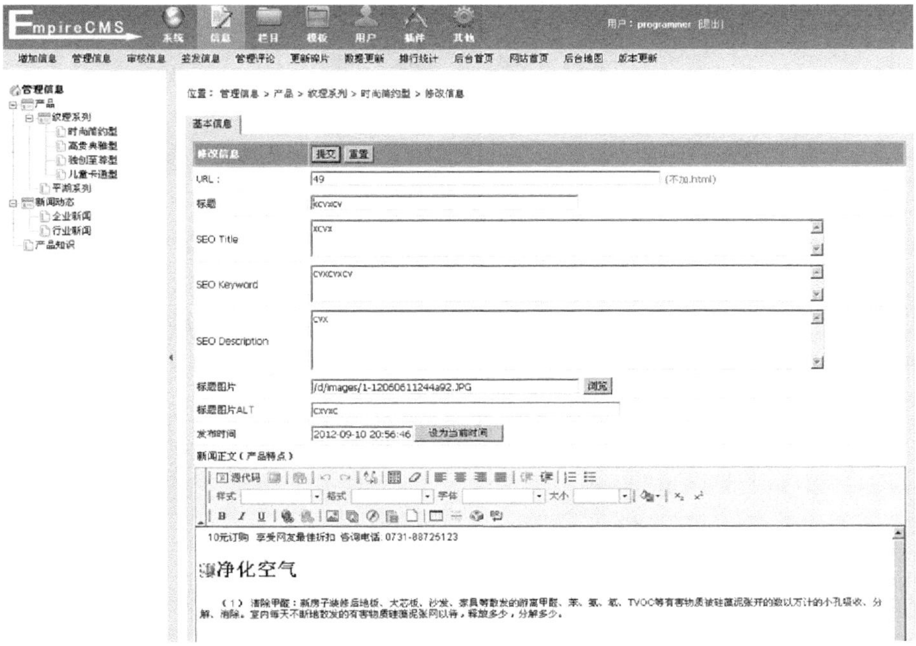

图 14.38　添加新闻

图 14.39　浏览新闻

(4) 模板模块。(30 分)

(5) 栏目模块。(20 分)

(6) 信息模块。(10 分)

(7) 创新性、其他功能。(10 分)

参 考 文 献

[1] 徐辉. PHPWeb 程序设计教程与实验. 北京:清华大学出版社,2008.
[2] 李晓斌. PHP＋MySQL＋Dreamweaver 网站建设全程揭秘. 北京:清华大学出版社,2014.
[3] 软件开发技术联盟. PHP＋MySQL 开发实战. 北京:清华大学出版社,2013.
[4] 唐四薪. PHPWeb 程序设计与 Ajax 技术. 北京:清华大学出版社,2014.
[5] 于荷云. PHP＋MySQL 网站开发全程实例. 北京:清华大学出版社,2012.
[6] 郑阿奇. MySQL 实用教程. 北京:电子工业出版社,2014.
[7] 郑阿奇. PHP 实用教程. 北京:电子工业出版社,2014.
[8] 王甲临. PHP 程序设计经典 300 例. 北京:电子工业出版社,2013.
[9] jQuery 教程. http://www.w3school.com.cn/jquery/.